# Craig's Soil Mechanics Seventh Edition
# Solutions Manual

D0222532

# Craig's Soil Mechanics Seventh Edition Solutions Manual

## R.F. Craig

*Formerly*
*Department of Civil Engineering*
*University of Dundee UK*

Taylor & Francis
Taylor & Francis Group

LONDON AND NEW YORK

First published 1992
by Taylor & Francis
Second edition 1997
Third edition 2004
2 Park Square, Milton Park, Abingdon, Oxon, OX14 4RN

Simultaneously published in the USA and Canada
by Taylor & Francis
270 Madison Ave, New York NY 10016

*Taylor & Francis is an imprint of the Taylor & Francis Group*

Transferred to Digital Printing 2006

© 1992, 1997, 2004 R.F. Craig

Typeset in 10/12pt Times by
Integra Software Services Pvt. Ltd, Pondicherry, India

*British Library Cataloguing in Publication Data*
A catalogue record for this book is available
from the British Library

*Library of Congress Cataloging in Publication Data*
A catalog record for this book has been requested

ISBN 0–415–33294–X

# Contents

1   Basic characteristics of soils    1

2   Seepage    6

3   Effective stress    14

4   Shear strength    22

5   Stresses and displacements    28

6   Lateral earth pressure    34

7   Consolidation theory    50

8   Bearing capacity    60

9   Stability of slopes    74

# Chapter 1

# Basic characteristics of soils

## 1.1

Soil E consists of 98% coarse material (31% gravel size; 67% sand size) and 2% fines. It is classified as SW: well-graded gravelly SAND or, in greater detail, well-graded slightly silty very gravelly SAND.

Soil F consists of 63% coarse material (2% gravel size; 61% sand size) and 37% non-plastic fines (i.e. between 35 and 65% fines); therefore, the soil is classified as MS: sandy SILT.

Soil G consists of 73% fine material (i.e. between 65 and 100% fines) and 27% sand size. The liquid limit is 32 and the plasticity index is 8 (i.e. 32 − 24), plotting marginally below the A-line in the ML zone on the plasticity chart. Thus the classification is ML: SILT (M-SOIL) of low plasticity. (The plasticity chart is given in Figure 1.7.)

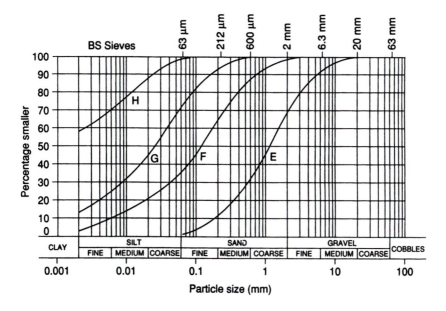

Figure Q1.1

Soil H consists of 99% fine material (58% clay size; 47% silt size). The liquid limit is 78 and the plasticity index is 47 (i.e. 78 – 31), plotting above the A-line in the CV zone on the plasticity chart. Thus the classification is CV: CLAY of very high plasticity.

## 1.2

From Equation 1.17

$$1 + e = G_s(1 + w)\frac{\rho_w}{\rho} = 2.70 \times 1.095 \times \frac{1.00}{1.91} = 1.55$$

$$\therefore e = 0.55$$

Using Equation 1.13

$$S_r = \frac{wG_s}{e} = \frac{0.095 \times 2.70}{0.55} = 0.466 \quad (46.6\%)$$

Using Equation 1.19

$$\rho_{sat} = \frac{G_s + e}{1 + e}\rho_w = \frac{3.25}{1.55} \times 1.00 = 2.10\,\text{Mg/m}^3$$

From Equation 1.14

$$w = \frac{e}{G_s} = \frac{0.55}{2.70} = 0.204 \quad (20.4\%)$$

## 1.3

Equations similar to 1.17–1.20 apply in the case of unit weights; thus,

$$\gamma_d = \frac{G_s}{1 + e}\gamma_w = \frac{2.72}{1.70} \times 9.8 = 15.7\,\text{kN/m}^3$$

$$\gamma_{sat} = \frac{G_s + e}{1 + e}\gamma_w = \frac{3.42}{1.70} \times 9.8 = 19.7\,\text{kN/m}^3$$

Using Equation 1.21

$$\gamma' = \frac{G_s - 1}{1 + e}\gamma_w = \frac{1.72}{1.70} \times 9.8 = 9.9\,\text{kN/m}^3$$

Using Equation 1.18a with $S_r = 0.75$

$$\gamma = \frac{G_s + S_r e}{1 + e}\gamma_w = \frac{3.245}{1.70} \times 9.8 = 18.7\,\text{kN/m}^3$$

Using Equation 1.13

$$w = \frac{S_r e}{G_s} = \frac{0.75 \times 0.70}{2.72} = 0.193 \quad (19.3\%)$$

The reader should not attempt to memorize the above equations. Figure 1.10(b) should be drawn and, from a knowledge of the definitions, relevant expressions can be written by inspection.

## 1.4

$$\text{Volume of specimen} = \frac{\pi}{4} \times 38^2 \times 76 = 86\,200 \text{ mm}^3$$

$$\text{Bulk density } (\rho) = \frac{\text{Mass}}{\text{Volume}} = \frac{168.0}{86\,200 \times 10^{-3}} = 1.95 \text{ Mg/m}^3$$

$$\text{Water content } (w) = \frac{168.0 - 130.5}{130.5} = 0.287 \quad (28.7\%)$$

From Equation 1.17

$$1 + e = G_s(1 + w)\frac{\rho_w}{\rho} = 2.73 \times 1.287 \times \frac{1.00}{1.95} = 1.80$$

$$\therefore e = 0.80$$

Using Equation 1.13

$$S_r = \frac{wG_s}{e} = \frac{0.287 \times 2.73}{0.80} = 0.98 \quad (98\%)$$

## 1.5

Using Equation 1.24

$$\rho_d = \frac{\rho}{1 + w} = \frac{2.15}{1.12} = 1.92 \text{ Mg/m}^3$$

From Equation 1.17

$$1 + e = G_s(1 + w)\frac{\rho_w}{\rho} = 2.65 \times 1.12 \times \frac{1.00}{2.15} = 1.38$$

$$\therefore e = 0.38$$

Using Equation 1.13

$$S_r = \frac{wG_s}{e} = \frac{0.12 \times 2.65}{0.38} = 0.837 \quad (83.7\%)$$

Using Equation 1.15

$$A = \frac{e - wG_s}{1 + e} = \frac{0.38 - 0.318}{1.38} = 0.045 \quad (4.5\%)$$

The zero air voids dry density is given by Equation 1.25

$$\rho_d = \frac{G_s}{1 + wG_s}\rho w = \frac{2.65}{1 + (0.135 \times 2.65)} \times 1.00 = 1.95\,\mathrm{Mg/m^3}$$

i.e. a dry density of $2.00\,\mathrm{Mg/m^3}$ would not be possible.

### 1.6

| Mass (g) | $\rho$ (Mg/m³) | $w$ | $\rho_d$ (Mg/m³) | $\rho_{d_0}$ (Mg/m³) | $\rho_{d_5}$ (Mg/m³) | $\rho_{d_{10}}$ (Mg/m³) |
|---|---|---|---|---|---|---|
| 2010 | 2.010 | 0.128 | 1.782 | 1.990 | 1.890 | 1.791 |
| 2092 | 2.092 | 0.145 | 1.827 | 1.925 | 1.829 | 1.733 |
| 2114 | 2.114 | 0.156 | 1.829 | 1.884 | 1.790 | 1.696 |
| 2100 | 2.100 | 0.168 | 1.798 | 1.843 | 1.751 | 1.658 |
| 2055 | 2.055 | 0.192 | 1.724 | 1.765 | 1.676 | 1.588 |

In each case the bulk density ($\rho$) is equal to the mass of compacted soil divided by the volume of the mould. The corresponding value of dry density ($\rho_d$) is obtained from Equation 1.24. The dry density–water content curve is plotted, from which

$$w_{opt} = 15\% \quad \text{and} \quad \rho_{d_{max}} = 1.83\,\mathrm{Mg/m^3}$$

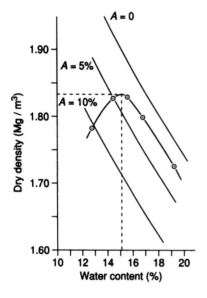

Figure Q1.6

Equation 1.26, with $A$ equal, in turn, to 0, 0.05 and 0.10, is used to calculate values of dry density ($\rho_{d_0}$, $\rho_{d_5}$, $\rho_{d_{10}}$ respectively) for use in plotting the air content curves. The experimental values of $w$ have been used in these calculations; however, any series of $w$ values within the relevant range could be used. By inspection, the value of air content at maximum dry density is 3.5%.

## 1.7

From Equation 1.20

$$e = \frac{G_s \rho_w}{\rho_d} - 1$$

The maximum and minimum values of void ratio are given by

$$e_{max} = \frac{G_s \rho w}{\rho_{d_{min}}} - 1$$

$$e_{min} = \frac{G_s \rho_w}{\rho_{d_{max}}} - 1$$

From Equation 1.23

$$I_D = \frac{G_s \rho_w (1/\rho_{d_{min}} - 1/\rho_d)}{G_s \rho_w (1/\rho_{d_{min}} - 1/\rho_{d_{max}})}$$

$$= \frac{[1 - (\rho_{d_{min}}/\rho_d)]1/\rho_{d_{min}}}{[1 - (\rho_{d_{min}}/\rho_{d_{max}})]1/\rho_{d_{min}}}$$

$$= \left(\frac{\rho_d - \rho_{d_{min}}}{\rho_{d_{max}} - \rho_{d_{min}}}\right)\frac{\rho_{d_{max}}}{\rho_d}$$

$$= \left(\frac{1.72 - 1.54}{1.81 - 1.54}\right)\frac{1.81}{1.72}$$

$$= 0.70 \quad (70\%)$$

# Chapter 2

# Seepage

## 2.1

The coefficient of permeability is determined from the equation

$$k = 2.3 \frac{al}{At_1} \log \frac{h_0}{h_1}$$

where

$$a = \frac{\pi}{4} \times 0.005^2 \, \mathrm{m}^2, \qquad l = 0.2 \, \mathrm{m}$$

$$A = \frac{\pi}{4} \times 0.1^2 \, \mathrm{m}^2, \qquad t_1 = 3 \times 60^2 \, \mathrm{s}$$

$$\log \frac{h_0}{h_1} = \log \frac{1.00}{0.35} = 0.456$$

$$\therefore k = \frac{2.3 \times 0.005^2 \times 0.2 \times 0.456}{0.1^2 \times 3 \times 60^2} = 4.9 \times 10^{-8} \, \mathrm{m/s}$$

## 2.2

The flow net is drawn in Figure Q2.2. In the flow net there are 3.7 flow channels and 11 equipotential drops, i.e. $N_f = 3.7$ and $N_d = 11$. The overall loss in total head is 4.00 m. The quantity of seepage is calculated by using Equation 2.16:

$$q = kh \frac{N_f}{N_d} = 10^{-6} \times 4.00 \times \frac{3.7}{11} = 1.3 \times 10^{-6} \, \mathrm{m^3/s} \text{ per m}$$

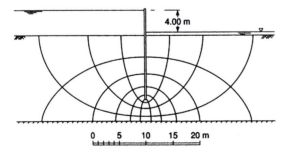

4.00 m

0   5   10   15   20 m

Figure Q2.2

**2.3**

The flow net is drawn in Figure Q2.3, from which $N_f = 3.5$ and $N_d = 9$. The overall loss in total head is 3.00 m. Then,

$$q = kh\frac{N_f}{N_d} = 5 \times 10^{-5} \times 3.00 \times \frac{3.5}{9} = 5.8 \times 10^{-5}\,\text{m}^3/\text{s per m}$$

The pore water pressure is determined at the points of intersection of the equipotentials with the base of the structure. The total head ($h$) at each point is obtained from the flow net. The elevation head ($z$) at each point on the base of the structure is $-2.50$ m. The calculations are tabulated below and the distribution of pressure ($u$) is plotted to scale in the figure.

| Point | $h$ (m) | $h - z$ (m) | $u = \gamma_w(h - z)$ (kN/m²) |
|---|---|---|---|
| 1 | 2.33 | 4.83 | 47 |
| 2 | 2.00 | 4.50 | 44 |
| 3 | 1.67 | 4.17 | 41 |
| 4 | 1.33 | 3.83 | 37 |
| 5 | 1.00 | 3.50 | 34 |
| 6 | 0.67 | 3.17 | 31 |

e.g. for Point 1:

$$h_1 = \frac{7}{9} \times 3.00 = 2.33\,\text{m}$$

$$h_1 - z_1 = 2.33 - (-2.50) = 4.83\,\text{m}$$

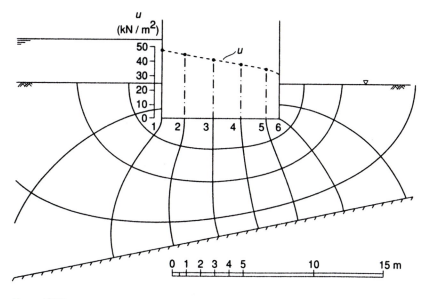

Figure Q2.3

$$u_1 = 9.8 \times 4.83 = 47 \, \text{kN/m}^2$$

The uplift force on the base of the structure is equal to the area of the pressure diagram and is 316 kN per unit length.

## 2.4

The flow net is drawn in Figure Q2.4, from which $N_f = 10.0$ and $N_d = 11$. The overall loss in total head is 5.50 m. Then,

$$q = kh\frac{N_f}{N_d} = 4.0 \times 10^{-7} \times 5.50 \times \frac{10}{11} = 2.0 \times 10^{-6} \, \text{m}^3/\text{s per m}$$

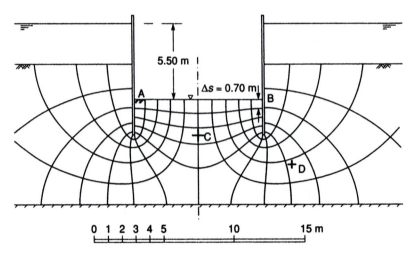

Figure Q2.4

## 2.5

The flow net is drawn in Figure Q2.5, from which $N_f = 4.2$ and $N_d = 9$. The overall loss in total head is 5.00 m. Then,

$$q = kh\frac{N_f}{N_d} = 2.0 \times 10^{-6} \times 5.00 \times \frac{4.2}{9} = 4.7 \times 10^{-6} \, \text{m}^3/\text{s per m}$$

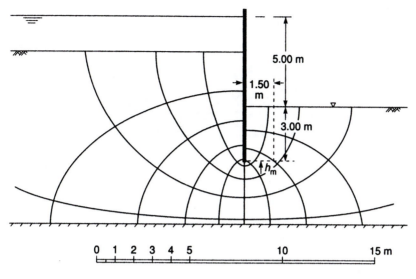

5.00 m

1.50 m

3.00 m

$h_m$

0 1 2 3 4 5        10        15 m

*Figure Q2.5*

## 2.6

The scale transformation factor in the $x$ direction is given by Equation 2.21, i.e.

$$x_t = x \frac{\sqrt{k_z}}{\sqrt{k_x}} = x \frac{\sqrt{1.8}}{\sqrt{5.0}} = 0.60x$$

Thus in the transformed section the horizontal dimension 33.00 m becomes (33.00 × 0.60), i.e. 19.80 m, and the slope 1:5 becomes 1:3. All dimensions in the vertical direction are unchanged. The transformed section is shown in Figure Q2.6 and the flow net is drawn as for the isotropic case. From the flow net, $N_f = 3.25$ and $N_d = 12$. The overall loss in total head is 14.00 m. The equivalent isotropic permeability applying to the transformed section is given by Equation 2.23, i.e.

$$k' = \sqrt{(k_x k_z)} = \sqrt{(5.0 \times 1.8)} \times 10^{-7} = 3.0 \times 10^{-7}\,\text{m/s}$$

Thus the quantity of seepage is given by

$$q = k'h\frac{N_f}{N_d} = 3.0 \times 10^{-7} \times 14.00 \times \frac{3.25}{12} = 1.1 \times 10^{-6}\,\text{m}^3/\text{s per m}$$

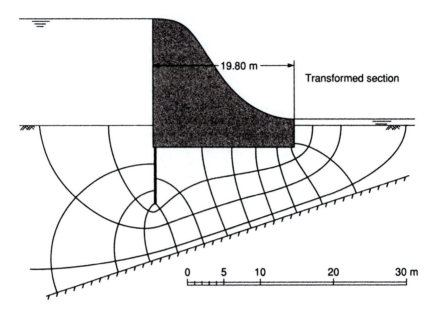

*Figure Q2.6*

## 2.7

The scale transformation factor in the $x$ direction is

$$x_t = x\frac{\sqrt{k_z}}{\sqrt{k_x}} = x\frac{\sqrt{2.7}}{\sqrt{7.5}} = 0.60x$$

Thus all dimensions in the $x$ direction are multipled by 0.60. All dimensions in the $z$ direction are unchanged. The transformed section is shown in Figure Q2.7. The equivalent isotropic permeability is

$$k' = \sqrt{(k_x k_z)} = \sqrt{(7.5 \times 2.7)} \times 10^{-6} = 4.5 \times 10^{-6}\,\text{m/s}$$

The focus of the basic parabola is at point A. The parabola passes through point G such that

$$GC = 0.3\,HC = 0.3 \times 30 = 9.0\,\text{m}$$

Thus the coordinates of G are

$$x = -48.0 \quad \text{and} \quad z = +20.0$$

Substituting these coordinates in Equation 2.34

$$-48.0 = x_0 - \frac{20.0^2}{4x_0}$$

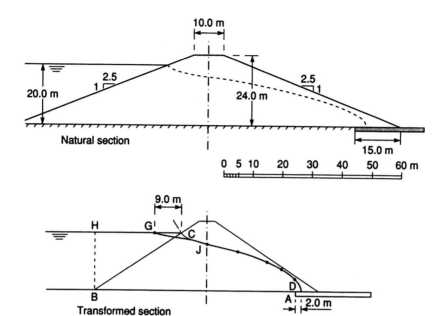

Figure Q2.7

Hence,

$$x_0 = 2.0\,\text{m}$$

Using Equation 2.34, with $x_0 = 2.0\,\text{m}$, the coordinates of a number of points on the basic parabola are calculated, i.e.

$$x = 2.0 - \frac{z^2}{8.0}$$

| x | 2.0 | 0 | −5.0 | −10.0 | −20.0 | −30.0 |
|---|-----|-----|------|-------|-------|-------|
| z | 0 | 4.00 | 7.48 | 9.80 | 13.27 | 16.00 |

The basic parabola is plotted in Figure Q2.7. The upstream correction is drawn using personal judgement.

No downstream correction is required in this case since $\beta = 180°$. If required, the top flow line can be plotted back onto the natural section, the $x$ coordinates above being divided by the scale transformation factor. The quantity of seepage can be calculated using Equation 2.33, i.e.

$$q = 2k'x_0 = 2 \times 4.5 \times 10^{-6} \times 2.0 = 1.8 \times 10^{-5}\,\text{m}^3/\text{s per m}$$

## 2.8

The flow net is drawn in Figure Q2.8, from which $N_f = 3.3$ and $N_d = 7$. The overall loss in total head is 2.8 m. Then,

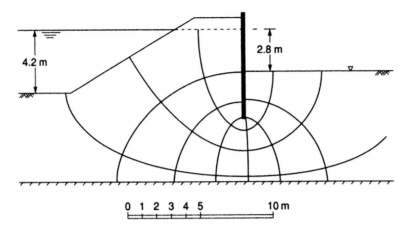

**Figure Q2.8**

$$q = kh\frac{N_f}{N_d} = 4.5 \times 10^{-5} \times 2.8 \times \frac{3.3}{7}$$
$$= 5.9 \times 10^{-5} \, \text{m}^3/\text{s per m}$$

## 2.9

The two isotropic soil layers, each 5 m thick, can be considered as a single homogeneous anisotropic layer of thickness 10 m in which the coefficients of permeability in the horizontal and vertical directions, respectively, are given by Equations 2.24 and 2.25, i.e.

$$\bar{k}_x = \frac{H_1 k_1 + H_2 k_2}{H_1 + H_2} = \frac{10^{-6}}{10}\{(5 \times 2.0) + (5 \times 16)\} = 9.0 \times 10^{-6} \, \text{m/s}$$

$$\bar{k}_z = \frac{H_1 + H_2}{\dfrac{H_1}{k_1} + \dfrac{H_2}{k_2}} = \frac{10}{\dfrac{5}{(2 \times 10^{-6})} + \dfrac{5}{(16 \times 10^{-6})}}$$
$$= 3.6 \times 10^{-6} \, \text{m/s}$$

Then the scale transformation factor is given by

$$x_t = x\frac{\sqrt{\bar{k}_z}}{\sqrt{\bar{k}_x}} = x\frac{\sqrt{3.6}}{\sqrt{9.0}} = 0.63x$$

Thus in the transformed section the dimension 10.00 m becomes 6.30 m; vertical dimensions are unchanged. The transformed section is shown in Figure Q2.9 and the flow net is drawn as for a single isotropic layer. From the flow net, $N_f = 5.6$ and $N_d = 11$. The overall loss in total head is 3.50 m. The equivalent isotropic permeability is

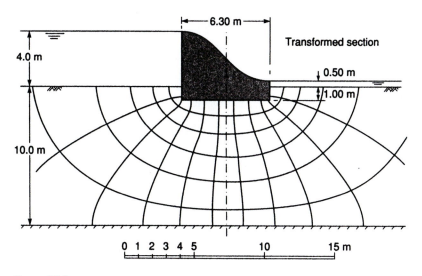

*Figure  Q2.9*

$$k' = \sqrt{(\bar{k}_x \bar{k}_z)} = \sqrt{(9.0 \times 3.6)} \times 10^{-6} = 5.7 \times 10^{-6} \, \text{m/s}$$

Then the quantity of seepage is given by

$$q = k'h \frac{N_f}{N_d} = 5.7 \times 10^{-6} \times 3.50 \times \frac{5.6}{11}$$
$$= 1.0 \times 10^{-5} \, \text{m}^3/\text{s per m}$$

# Chapter 3

# Effective stress

## 3.1

Buoyant unit weight:

$$\gamma' = \gamma_{\text{sat}} - \gamma_\text{w} = 20 - 9.8 = 10.2\,\text{kN/m}^3$$

Effective vertical stress:

$$\sigma'_\text{v} = 5 \times 10.2 = 51\,\text{kN/m}^2 \quad \textbf{or}$$

Total vertical stress:

$$\sigma_\text{v} = (2 \times 9.8) + (5 \times 20) = 119.6\,\text{kN/m}^2$$

Pore water pressure:

$$u = 7 \times 9.8 = 68.6\,\text{kN/m}^2$$

Effective vertical stress:

$$\sigma'_\text{v} = \sigma_\text{v} - u = 119.6 - 68.6 = 51\,\text{kN/m}^2$$

## 3.2

Buoyant unit weight:

$$\gamma' = \gamma_{\text{sat}} - \gamma_\text{w} = 20 - 9.8 = 10.2\,\text{kN/m}^3$$

Effective vertical stress:

$$\sigma'_\text{v} = 5 \times 10.2 = 51\,\text{kN/m}^2 \quad \textbf{or}$$

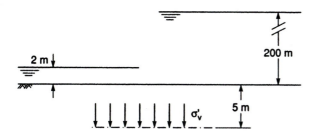

Figure Q3.1/3.2

Total vertical stress:

$$\sigma_v = (200 \times 9.8) + (5 \times 20) = 2060 \, \text{kN/m}^2$$

Pore water pressure:

$$u = 205 \times 9.8 = 2009 \, \text{kN/m}^2$$

Effective vertical stress:

$$\sigma_v' = \sigma_v - u = 2060 - 2009 = 51 \, \text{kN/m}^2$$

## 3.3

At top of the clay:

$$\sigma_v = (2 \times 16.5) + (2 \times 19) = 71.0 \, \text{kN/m}^2$$
$$u = 2 \times 9.8 = 19.6 \, \text{kN/m}^2$$
$$\sigma_v' = \sigma_v - u = 71.0 - 19.6 = 51.4 \, \text{kN/m}^2$$

Alternatively,

$$\gamma' \, (\text{sand}) = 19 - 9.8 = 9.2 \, \text{kN/m}^3$$
$$\sigma_v' = (2 \times 16.5) + (2 \times 9.2) = 51.4 \, \text{kN/m}^2$$

At bottom of the clay:

$$\sigma_v = (2 \times 16.5) + (2 \times 19) + (4 \times 20) = 151.0 \, \text{kN/m}^2$$
$$u = 12 \times 9.8 = 117.6 \, \text{kN/m}^2$$
$$\sigma_v' = \sigma_v - u = 151.0 - 117.6 = 33.4 \, \text{kN/m}^2$$

NB   The alternative method of calculation is not applicable because of the artesian condition.

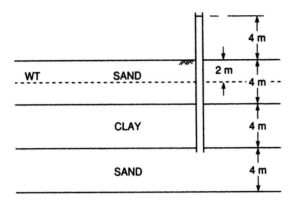

Figure Q3.3

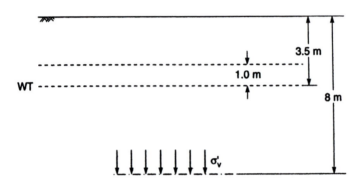

Figure Q3.4

**3.4**

$$\gamma' = 20 - 9.8 = 10.2\,\text{kN/m}^3$$

At 8 m depth:

$$\sigma'_v = (2.5 \times 16) + (1.0 \times 20) + (4.5 \times 10.2) = 105.9\,\text{kN/m}^2$$

**3.5**

$$\gamma'\,(\text{sand}) = 19 - 9.8 = 9.2\,\text{kN/m}^3$$
$$\gamma'\,(\text{clay}) = 20 - 9.8 = 10.2\,\text{kN/m}^3$$

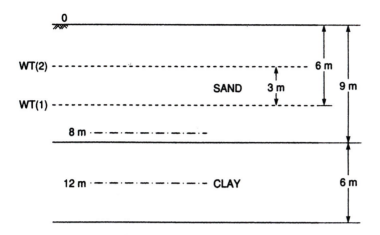

*Figure Q3.5*

(a) Immediately after WT rise:

At 8 m depth, pore water pressure is governed by the new WT level because the permeability of the sand is high.

$$\therefore \sigma'_v = (3 \times 16) + (5 \times 9.2) = 94.0 \, \text{kN/m}^2$$

At 12 m depth, pore water pressure is governed by the old WT level because the permeability of the clay is very low. (However, there will be an increase in total stress of 9 kN/m² due to the increase in unit weight from 16 to 19 kN/m² between 3 and 6 m depth: this is accompanied by an immediate increase of 9 kN/m² in pore water pressure.)

$$\therefore \sigma'_v = (6 \times 16) + (3 \times 9.2) + (3 \times 10.2) = 154.2 \, \text{kN/m}^2$$

(b) Several years after WT rise:

At both depths, pore water pressure is governed by the new WT level, it being assumed that swelling of the clay is complete.
  At 8 m depth:

$$\sigma'_v = 94.0 \, \text{kN/m}^2 \quad \text{(as above)}$$

At 12 m depth:

$$\sigma'_v = (3 \times 16) + (6 \times 9.2) + (3 \times 10.2) = 133.8 \, \text{kN/m}^2$$

## 3.6

Total weight:

$$\overline{ab} = 21.0\,\text{kN}$$

Effective weight:

$$\overline{ac} = 11.2\,\text{kN}$$

Resultant boundary water force:

$$\overline{be} = 11.9\,\text{kN}$$

Seepage force:

$$\overline{ce} = 3.4\,\text{kN}$$

Resultant body force:

$$\overline{ae} = 9.9\,\text{kN}\ (73°\ \text{to horizontal})$$

(Refer to Figure Q3.6.)

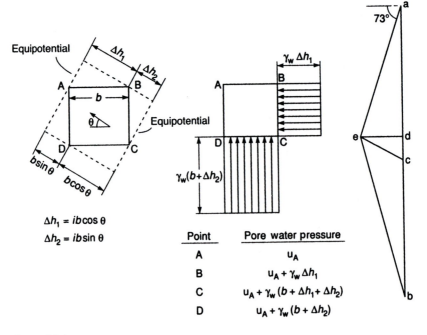

| Point | Pore water pressure |
|-------|---------------------|
| A | $u_A$ |
| B | $u_A + \gamma_w\,\Delta h_1$ |
| C | $u_A + \gamma_w\,(b + \Delta h_1 + \Delta h_2)$ |
| D | $u_A + \gamma_w\,(b + \Delta h_2)$ |

Figure Q3.6

## 3.7

Situation (1):
(a)

$$\sigma = 3\gamma_w + 2\gamma_{sat} = (3 \times 9.8) + (2 \times 20) = 69.4\,\text{kN/m}^2$$
$$u = \gamma_w(h - z) = 9.8\{1 - (-3)\} = 39.2\,\text{kN/m}^2$$
$$\sigma' = \sigma - u = 69.4 - 39.2 = 30.2\,\text{kN/m}^2$$

(b)

$$i = \frac{2}{4} = 0.5$$
$$j = i\gamma_w = 0.5 \times 9.8 = 4.9\,\text{kN/m}^3 \downarrow$$
$$\sigma' = 2(\gamma' + j) = 2(10.2 + 4.9) = 30.2\,\text{kN/m}^2$$

Situation (2):
(a)

$$\sigma = 1\gamma_w + 2\gamma_{sat} = (1 \times 9.8) + (2 \times 20) = 49.8\,\text{kN/m}^2$$
$$u = \gamma_w(h - z) = 9.8\{1 - (-3)\} = 39.2\,\text{kN/m}^2$$
$$\sigma' = \sigma - u = 49.8 - 39.2 = 10.6\,\text{kN/m}^2$$

(b)

$$i = \frac{2}{4} = 0.5$$
$$j = i\gamma_w = 0.5 \times 9.8 = 4.9\,\text{kN/m}^3 \uparrow$$
$$\sigma' = 2(\gamma' - j) = 2(10.2 - 4.9) = 10.6\,\text{kN/m}^2$$

## 3.8

The flow net is drawn in Figure Q2.4.

Loss in total head between adjacent equipotentials:

$$\Delta h = \frac{5.50}{N_d} = \frac{5.50}{11} = 0.50\,\text{m}$$

Exit hydraulic gradient:

$$i_e = \frac{\Delta h}{\Delta s} = \frac{0.50}{0.70} = 0.71$$

The critical hydraulic gradient is given by Equation 3.9:

$$i_c = \frac{\gamma'}{\gamma_w} = \frac{10.2}{9.8} = 1.04$$

Therefore, factor of safety against 'boiling' (Equation 3.11):

$$F = \frac{i_c}{i_e} = \frac{1.04}{0.71} = 1.5$$

Total head at C:

$$h_C = \frac{n_d}{N_d}h = \frac{2.4}{11} \times 5.50 = 1.20\,\text{m}$$

Elevation head at C:

$$z_C = -2.50\,\text{m}$$

Pore water pressure at C:

$$u_C = 9.8(1.20 + 2.50) = 36\,\text{kN/m}^2$$

Therefore, effective vertical stress at C:

$$\sigma'_C = \sigma_C - u_C = (2.5 \times 20) - 36 = 14\,\text{kN/m}^2$$

For point D:

$$h_D = \frac{7.3}{11} \times 5.50 = 3.65\,\text{m}$$

$$z_D = -4.50\,\text{m}$$

$$u_D = 9.8(3.65 + 4.50) = 80\,\text{kN/m}^2$$

$$\sigma'_D = \sigma_D - u_D = (3 \times 9.8) + (7 \times 20) - 80 = 90\,\text{kN/m}^2$$

**3.9**

The flow net is drawn in Figure Q2.5.

For a soil prism 1.50 × 3.00 m, adjacent to the piling:

$$h_m = \frac{2.6}{9} \times 5.00 = 1.45\,\text{m}$$

Factor of safety against 'heaving' (Equation 3.10):

$$F = \frac{i_c}{i_m} = \frac{\gamma' d}{\gamma_w h_m} = \frac{9.7 \times 3.00}{9.8 \times 1.45} = 2.0$$

With a filter:

$$F = \frac{\gamma' d + w}{\gamma_w h_m}$$

$$\therefore 3 = \frac{(9.7 \times 3.00) + w}{9.8 \times 1.45}$$

$$\therefore w = 13.5 \, \text{kN/m}^2$$

Depth of filter $= 13.5/21 = 0.65 \, \text{m}$ (if above water level).

# Chapter 4

# Shear strength

## 4.1

$$\sigma = 295\,\text{kN/m}^2$$
$$u = 120\,\text{kN/m}^2$$
$$\sigma' = \sigma - u = 295 - 120 = 175\,\text{kN/m}^2$$
$$\tau_f = c' + \sigma' \tan \phi' = 12 + 175 \tan 30° = 113\,\text{kN/m}^2$$

## 4.2

| $\sigma_3'$ (kN/m²) | $\sigma_1 - \sigma_3$ (kN/m²) | $\sigma_1'$ (kN/m²) |
|---|---|---|
| 100 | 452 | 552 |
| 200 | 908 | 1108 |
| 400 | 1810 | 2210 |
| 800 | 3624 | 4424 |

The Mohr circles are drawn in Figure Q4.2, together with the failure envelope from which $\phi' = 44°$.

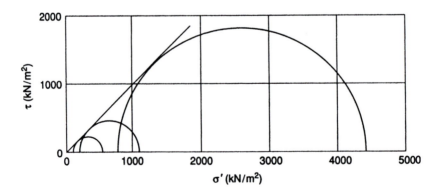

Figure Q4.2

**4.3**

| $\sigma_3$ (kN/m$^2$) | $\sigma_1 - \sigma_3$ (kN/m$^2$) | $\sigma_1$ (kN/m$^2$) |
|---|---|---|
| 200 | 222 | 422 |
| 400 | 218 | 618 |
| 600 | 220 | 820 |

The Mohr circles and failure envelope are drawn in Figure Q4.3, from which $c_u = 110\,\text{kN/m}^2$ and $\phi_u = 0$.

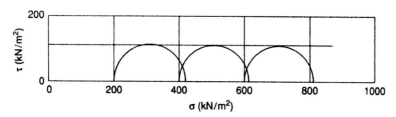

*Figure Q4.3*

**4.4**

The modified shear strength parameters are

$$\alpha' = \tan^{-1}(\sin \phi') = \tan^{-1}(\sin 29°) = 26°$$
$$a' = c' \cos \phi' = 15 \cos 29° = 13\,\text{kN/m}^2$$

The coordinates of the stress point representing failure conditions in the test are

$$\frac{1}{2}(\sigma_1 - \sigma_2) = \frac{1}{2} \times 170 = 85\,\text{kN/m}^2$$
$$\frac{1}{2}(\sigma_1 + \sigma_3) = \frac{1}{2}(270 + 100) = 185\,\text{kN/m}^2$$

The pore water pressure at failure is given by the horizontal distance between this stress point and the modified failure envelope. Thus from Figure Q4.4

$$u_f = 36\,\text{kN/m}^2$$

*Figure Q4.4*

**4.5**

| $\sigma_3$ (kN/m²) | $\sigma_1 - \sigma_3$ (kN/m²) | $\sigma_1$ (kN/m²) | $u$ (kN/m²) | $\sigma_3'$ (kN/m²) | $\sigma_1'$ (kN/m²) |
|---|---|---|---|---|---|
| 150 | 103 | 253 | 82 | 68 | 171 |
| 300 | 202 | 502 | 169 | 131 | 333 |
| 450 | 305 | 755 | 252 | 198 | 503 |
| 600 | 410 | 1010 | 331 | 269 | 679 |

The Mohr circles and failure envelope are drawn in Figure Q4.5, from which $c' = 0$ and $\phi' = 25\frac{1}{2}°$.

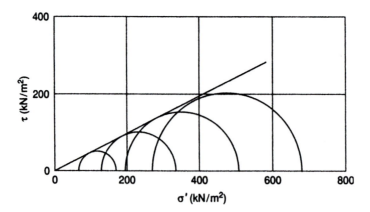

*Figure Q4.5*

The principal stress difference at failure depends only on the value of all-round pressure under which consolidation took place, i.e. 250 kN/m². Hence, by proportion, the expected value of $(\sigma_1 - \sigma_3)_f = 170\,\text{kN/m}^2$.

**4.6**

| $\sigma_3'$ (kN/m²) | $\Delta V/V_0$ | $\Delta l/l_0$ | Area (mm²) | Load (N) | $\sigma_1 - \sigma_3$ (kN/m²) | $\sigma_1'$ (kN/m²) |
|---|---|---|---|---|---|---|
| 200 | 0.061 | 0.095 | 1177 | 565 | 480 | 680 |
| 400 | 0.086 | 0.110 | 1165 | 1015 | 871 | 1271 |
| 600 | 0.108 | 0.124 | 1155 | 1321 | 1144 | 1744 |

The average cross-sectional area of each specimen is obtained from Equation 4.10; the original values of $A$, $l$ and $V$ are: $A_0 = 1134\,\text{mm}^2$, $l_0 = 76\,\text{mm}$, $V_0 = 86\,200\,\text{mm}^3$. The Mohr circles are drawn in Figure Q4.6(a) and (b). From (a) the secant parameters are measured as 34°, 31.5° and 29°. The failure envelope, shown in (b), exhibits a curvature and between 300 and 500 kN/m² is approximated to a straight line, from which $c' = 20\,\text{kN/m}^2$ and $\phi' = 31°$.

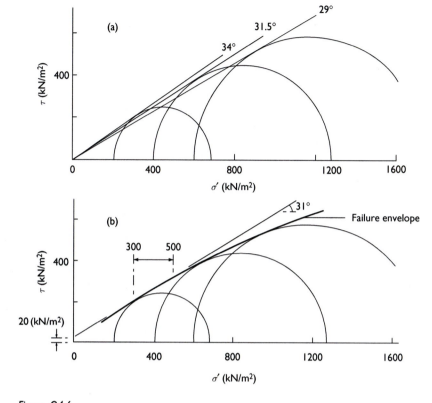

Figure Q4.6

## 4.7

The torque required to produce shear failure is given by

$$T = \pi d h \; c_u \frac{d}{2} + 2 \int_0^{d/2} 2\pi r \, dr c_u r$$

$$= \pi c_u \frac{d^2 h}{2} + 4\pi c_u \int_0^{d/2} r^2 dr$$

$$= \pi c_u \left( \frac{d^2 h}{2} + \frac{d^3}{6} \right)$$

Then,

$$35 = \pi c_u \left( \frac{5^2 \times 10}{2} + \frac{5^3}{6} \right) \times 10^{-3}$$

$$\therefore \; c_u = 76 \, \text{kN/m}^3$$

## 4.8

The relevant stress values are calculated as follows:

$\sigma_3 = 600\,\text{kN/m}^2$

| | | | | | | |
|---|---|---|---|---|---|---|
| $\sigma_1 - \sigma_3$ | 0 | 80 | 158 | 214 | 279 | 319 |
| $\sigma_1$ | 600 | 680 | 758 | 814 | 879 | 919 |
| $u$ | 200 | 229 | 277 | 318 | 388 | 433 |
| $\sigma_1'$ | 400 | 451 | 481 | 496 | 491 | 486 |
| $\sigma_3'$ | 400 | 371 | 323 | 282 | 212 | 167 |
| $\frac{1}{2}(\sigma_1 - \sigma_3)$ | 0 | 40 | 79 | 107 | 139 | 159 |
| $\frac{1}{2}(\sigma_1' + \sigma_3')$ | 400 | 411 | 402 | 389 | 351 | 326 |
| $\frac{1}{2}(\sigma_1 + \sigma_3)$ | 600 | 640 | 679 | 707 | 739 | 759 |

The stress paths are plotted in Figure Q4.8. The initial points on the effective and total stress paths are separated by the value of the back pressure ($u_s = 200\,\text{kN/m}^2$).

$$A_f = \frac{433 - 200}{319} = 0.73$$

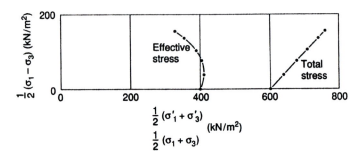

Figure Q4.8

## 4.9

$$B = \frac{\Delta u_3}{\Delta \sigma_3} = \frac{144}{350 - 200} = 0.96$$

| $\varepsilon_a$ (%) | $\Delta\sigma_1 = \sigma_1 - \sigma_3$ (kN/m²) | $\Delta u_1$ (kN/m²) | $\bar{A} = \Delta u_1 / \Delta\sigma_1$ |
|---|---|---|---|
| 0 | 0 | 0 | – |
| 2 | 201 | 100 | 0.50 |
| 4 | 252 | 96 | 0.38 |
| 6 | 275 | 78 | 0.28 |
| 8 | 282 | 68 | 0.24 |
| 10 | 283 | 65 | 0.23 |

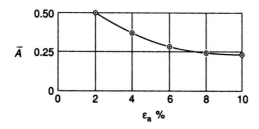

*Figure Q4.9*

The variation of $\overline{A}$ with axial strain is plotted in Figure Q4.9. At failure, $\overline{A} = 0.23$.

# Chapter 5

# Stresses and displacements

## 5.1

Vertical stress is given by

$$\sigma_z = \frac{Q}{z^2} I_\mathrm{p} = \frac{5000}{5^2} I_\mathrm{p}$$

Values of $I_\mathrm{p}$ are obtained from Table 5.1.

| r (m) | r/z | $I_\mathrm{p}$ | $\sigma_z$ (kN/m$^2$) |
|---|---|---|---|
| 0 | 0 | 0.478 | 96 |
| 1 | 0.2 | 0.433 | 87 |
| 2 | 0.4 | 0.329 | 66 |
| 3 | 0.6 | 0.221 | 44 |
| 4 | 0.8 | 0.139 | 28 |
| 5 | 1.0 | 0.084 | 17 |
| 7 | 1.4 | 0.032 | 6 |
| 10 | 2.0 | 0.009 | 2 |

The variation of $\sigma_z$ with radial distance ($r$) is plotted in Figure Q5.1.

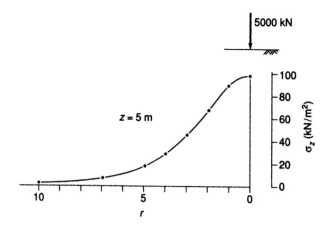

Figure Q5.1

**5.2**

Below the centre load (Figure Q5.2):

$$\frac{r}{z} = 0 \text{ for the 7500-kN load}$$

$$\therefore I_\mathrm{p} = 0.478$$

$$\frac{r}{z} = \frac{5}{4} = 1.25 \text{ for the 10 000- and 9000-kN loads}$$

$$\therefore I_\mathrm{p} = 0.045$$

Then,

$$\sigma_z = \sum \left( \frac{Q}{z^2} I_\mathrm{p} \right)$$

$$= \frac{7500 \times 0.478}{4^2} + \frac{10\,000 \times 0.045}{4^2} + \frac{9000 \times 0.045}{4^2}$$

$$= 224 + 28 + 25 = 277\,\mathrm{kN/m^2}$$

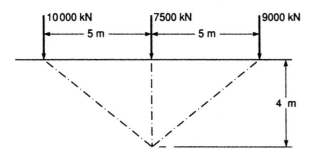

Figure Q5.2

**5.3**

The vertical stress under a corner of a rectangular area is given by

$$\sigma_z = qI_\mathrm{r}$$

where values of $I_\mathrm{r}$ are obtained from Figure 5.10. In this case

$$\sigma_z = 4 \times 250 \times I_\mathrm{r}\ (\mathrm{kN/m^2})$$

$$m = n = \frac{1}{z}$$

| z (m) | m, n | $I_r$ | $\sigma_z$ (kN/m²) |
|---|---|---|---|
| 0 | – | – | (250) |
| 0.5 | 2.00 | 0.233 | 233 |
| 1 | 1.00 | 0.176 | 176 |
| 1.5 | 0.67 | 0.122 | 122 |
| 2 | 0.50 | 0.085 | 85 |
| 3 | 0.33 | 0.045 | 45← |
| 4 | 0.25 | 0.027 | 27 |
| 7 | 0.14 | 0.009 | 9 |
| 10 | 0.10 | 0.005 | 5 |

$\sigma_z$ is plotted against $z$ in Figure Q5.3.

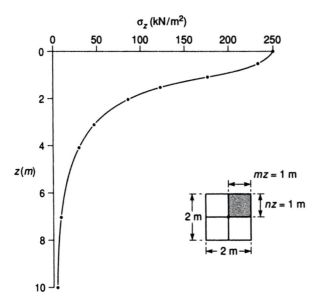

Figure Q5.3

**5.4**

(a)

$$m = \frac{12.5}{12} = 1.04$$

$$n = \frac{18}{12} = 1.50$$

From Figure 5.10, $I_r = 0.196$.

$$\therefore \ \sigma_z = 2 \times 175 \times 0.196 = 68 \, \text{kN/m}^2$$

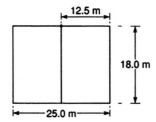

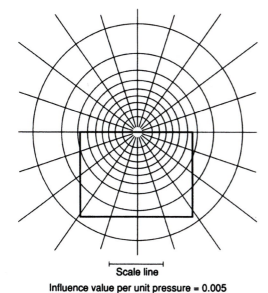

Scale line

**Influence value per unit pressure = 0.005**

*Figure Q5.4*

(b) The foundation is drawn on Newmark's chart as shown in Figure Q5.4, the scale line representing 12 m ($z$). Number of influence areas $(N) = 78$.

$$\therefore \sigma_z = 0.005 \, Nq = 0.005 \times 78 \times 175 = 68 \, \text{kN/m}^2$$

## 5.5

$Q = 150 \, \text{kN/m}$; $h = 4.00 \, \text{m}$; $m = 0.5$. The total thrust is given by Equation 5.18

$$P_x = \frac{2Q}{\pi} \frac{1}{m^2 + 1} = \frac{2 \times 150}{\pi \times 1.25} = 76 \, \text{kN/m}$$

Equation 5.17 is used to obtain the pressure distribution

$$p_x = \frac{4Q}{\pi h} \frac{m^2 n}{(m^2 + n^2)^2} = \frac{150}{\pi} \frac{m^2 n}{(m^2 + n^2)^2} \, (\text{kN/m}^2)$$

| $n$ | $\dfrac{m^2 n}{(m^2 + n^2)^2}$ | $p_x(\text{kN/m}^2)$ |
|-----|-------------------------------|----------------------|
| 0   | 0                             | 0                    |
| 0.1 | 0.370                         | 17.7                 |
| 0.2 | 0.595                         | 28.4                 |
| 0.3 | 0.649                         | 31.0                 |
| 0.4 | 0.595                         | 28.4                 |
| 0.6 | 0.403                         | 19.2                 |
| 0.8 | 0.252                         | 12.0                 |
| 1.0 | 0.160                         | 7.6                  |

The pressure distribution is plotted in Figure Q5.5.

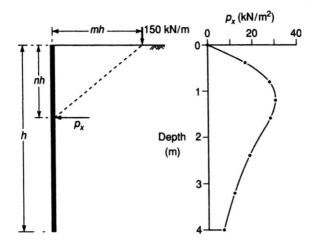

Figure Q5.5

## 5.6

$$\frac{H}{B} = \frac{10}{2} = 5$$

$$\frac{L}{B} = \frac{4}{2} = 2$$

$$\frac{D}{B} = \frac{1}{2} = 0.5$$

Hence from Figure 5.15

$$\mu_1 = 0.82$$

$$\mu_0 = 0.94$$

The immediate settlement is given by Equation 5.28

$$s_i = \mu_0 \mu_1 \frac{qB}{E_u}$$

$$= 0.94 \times 0.82 \times \frac{200 \times 2}{45} = 7\,\text{mm}$$

# Chapter 6

# Lateral earth pressure

## 6.1

For $\phi' = 37°$ the active pressure coefficient is given by

$$K_a = \frac{1 - \sin 37°}{1 + \sin 37°} = 0.25$$

The total active thrust (Equation 6.6a with $c' = 0$) is

$$P_a = \frac{1}{2} K_a \gamma H^2 = \frac{1}{2} \times 0.25 \times 17 \times 6^2 = 76.5 \, \text{kN/m}$$

If the wall is prevented from yielding, the at-rest condition applies. The approximate value of the coefficient of earth pressure at-rest is given by Equation 6.15a

$$K_0 = 1 - \sin \phi' = 1 - \sin 37° = 0.40$$

and the thrust on the wall is

$$P_0 = \frac{1}{2} K_0 \gamma H^2 = \frac{1}{2} \times 0.40 \times 17 \times 6^2 = 122 \, \text{kN/m}$$

## 6.2

The active pressure coefficients for the three soil types are as follows:

$$K_{a_1} = \frac{1 - \sin 35°}{1 + \sin 35°} = 0.271$$

$$K_{a_2} = \frac{1 - \sin 27°}{1 + \sin 27°} = 0.375, \qquad \sqrt{K_{a_2}} = 0.613$$

$$K_{a_3} = \frac{1 - \sin 42°}{1 + \sin 42°} = 0.198$$

Distribution of active pressure (plotted in Figure Q6.2):

| Depth (m) | Soil | Active pressure (kN/m$^2$) | |
|---|---|---|---|
| 3 | 1 | $0.271 \times 16 \times 3$ | $= 13.0$ |
| 5 | 1 | $(0.271 \times 16 \times 3) + (0.271 \times 9.2 \times 2)$ | $= 13.0 + 5.0 = 18.0$ |
| 5 | 2 | $\{(16 \times 3) + (9.2 \times 2)\}\ 0.375 - (2 \times 17 \times 0.613)$ | $= 24.9 - 20.9 = 4.0$ |
| 8 | 2 | $4.0 + (0.375 \times 10.2 \times 3)$ | $= 4.0 + 11.5 = 15.5$ |
| 8 | 3 | $\{(16 \times 3) + (9.2 \times 2) + (10.2 \times 3)\}\ 0.198$ | $= 19.2$ |
| 12 | 3 | $19.2 + (0.198 \times 11.2 \times 4)$ | $= 19.2 + 8.9 = 28.1$ |

At a depth of 12 m, the hydrostatic pressure $= 9.8 \times 9 = 88.2\,\text{kN/m}^2$. Calculation of total thrust and its point of application (forces are numbered as in Figure Q6.2 and moments are taken about the top of the wall) per m:

Total thrust $= 571\,\text{kN/m}$.

Point of application is (4893/571) m from the top of the wall, i.e. 8.57 m.

| Force (kN) | | Arm (m) | Moment (kN m) |
|---|---|---|---|
| (1) $\frac{1}{2} \times 0.271 \times 16 \times 3^2$ | $= 19.5$ | 2.0 | 39.0 |
| (2) $0.271 \times 16 \times 3 \times 2$ | $= 26.0$ | 4.0 | 104.0 |
| (3) $\frac{1}{2} \times 0.271 \times 9.2 \times 2^2$ | $= 5.0$ | 4.33 | 21.7 |
| (4) $[0.375 \{(16 \times 3) + (9.2 \times 2)\} - \{2 \times 17 \times 0.613\}]\ 3$ | $= 12.2$ | 6.5 | 79.3 |
| (5) $\frac{1}{2} \times 0.375 \times 10.2 \times 3^2$ | $= 17.2$ | 7.0 | 120.4 |
| (6) $[0.198 \{(16 \times 3) + (9.2 \times 2) + (10.2 \times 3)\}]\ 4$ | $= 76.8$ | 10.0 | 768.0 |
| (7) $\frac{1}{2} \times 0.198 \times 11.2 \times 4^2$ | $= 17.7$ | 10.67 | 188.9 |
| (8) $\frac{1}{2} \times 9.8 \times 9^2$ | $= 396.9$ | 9.0 | 3572.1 |
| | 571.3 | | 4893.4 |

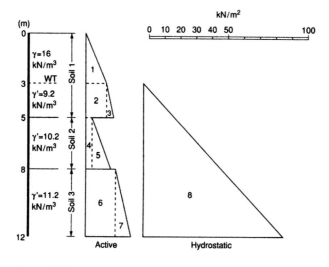

Figure Q6.2

## 6.3

(a) For $\phi_u = 0$, $K_a = K_p = 1$

$$K_{ac} = K_{pc} = 2\sqrt{1.5} = 2.45$$

At the lower end of the piling:

$$
\begin{aligned}
p_a &= K_a q + K_a \gamma_{sat} z - K_{ac} c_u \\
&= (1 \times 18 \times 3) + (1 \times 20 \times 4) - (2.45 \times 50) \\
&= 54 + 80 - 122.5 \\
&= 11.5 \, \text{kN/m}^2
\end{aligned}
$$

$$
\begin{aligned}
p_p &= K_p \gamma_{sat} z + K_{pc} c_u \\
&= (1 \times 20 \times 4) + (2.45 \times 50) \\
&= 80 + 122.5 \\
&= 202 \, \text{kN/m}^2
\end{aligned}
$$

(b) For $\phi' = 26°$ and $\delta = \dfrac{1}{2}\phi'$

$$K_a = 0.35$$
$$K_{ac} = 2\sqrt{(0.35 \times 1.5)} = 1.45 \quad \text{(Equation 6.19)}$$
$$K_p = 3.7$$
$$K_{pc} = 2\sqrt{(3.7 \times 1.5)} = 4.7 \quad \text{(Equation 6.24)}$$

At the lower end of the piling:

$$
\begin{aligned}
p_a &= K_a q + K_a \gamma' z - K_{ac} c' \\
&= (0.35 \times 18 \times 3) + (0.35 \times 10.2 \times 4) - (1.45 \times 10) \\
&= 18.9 + 14.3 - 14.5 \\
&= 18.7 \, \text{kN/m}^2
\end{aligned}
$$

$$
\begin{aligned}
p_p &= K_p \gamma' z + K_{pc} c' \\
&= (3.7 \times 10.2 \times 4) + (4.7 \times 10) \\
&= 151 + 47 \\
&= 198 \, \text{kN/m}^2
\end{aligned}
$$

**6.4**

(a) For $\phi' = 38°$, $K_a = 0.24$

$\gamma' = 20 - 9.8 = 10.2\,\text{kN/m}^3$

The pressure distribution is shown in Figure Q6.4. Consider moments (per m length of wall) about the toe.

| Force (kN) | | Arm (m) | Moment (kN m) |
|---|---|---|---|
| (1) $0.24 \times 10 \times 6.6$ | $= 15.9$ | 3.3 | 52.5 |
| (2) $\frac{1}{2} \times 0.24 \times 17 \times 3.9^2$ | $= 31.0$ | 4.00 | 124.0 |
| (3) $0.24 \times 17 \times 3.9 \times 2.7$ | $= 43.0$ | 1.35 | 58.0 |
| (4) $\frac{1}{2} \times 0.24 \times 10.2 \times 2.7^2$ | $= 8.9$ | 0.90 | 8.0 |
| (5) $\frac{1}{2} \times 9.8 \times 2.7^2$ | $= 35.7$ | 0.90 | 32.1 |
| | $H = 134.5$ | | $M_H = 274.6$ |
| (6) $6.2 \times 0.4 \times 23.5$ | $= 58.3$ | 1.20 | 70.0 |
| (7) $4.0 \times 0.4 \times 23.5$ | $= 37.6$ | 2.00 | 75.2 |
| (8) $3.9 \times 2.6 \times 17$ | $= 172.4$ | 2.70 | 465.5 |
| (9) $2.3 \times 2.6 \times 20$ | $= 119.6$ | 2.70 | 322.9 |
| (10) $100$ | $= 100.0$ | 1.20 | 120.0 |
| | $V = 487.9$ | | $M_V = 1053.6$ |

$$\sum M = M_V - M_H = 779.0\,\text{kN m}$$

Lever arm of base resultant:

$$\frac{\Sigma M}{V} = \frac{779}{488} = 1.60$$

Eccentricity of base resultant:

$$e = 2.00 - 1.60 = 0.40\,\text{m}$$

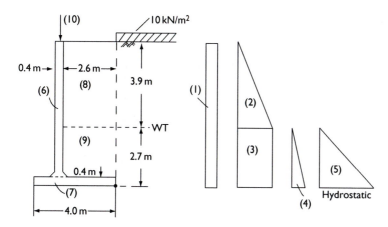

*Figure Q6.4*

Base pressures (Equation 6.27):

$$p = \frac{V}{B}\left(1 \pm \frac{6e}{B}\right)$$

$$= \frac{488}{4}(1 \pm 0.60)$$

$$= 195\,\text{kN/m}^2 \quad \text{and} \quad 49\,\text{kN/m}^2$$

Factor of safety against sliding (Equation 6.28):

$$F = \frac{V \tan \delta}{H} = \frac{488 \times \tan 25°}{134.5} = 1.7$$

(b) Using a partial factor of 1.25 the design value of $\phi'$ is $\tan^{-1}(\tan 38°/1.25) = 32°$. Therefore, $K_a = 0.31$ and the forces and moments are:

$$H = 163.3\,\text{kN}$$
$$V = 487.9\,\text{kN}$$
$$M_H = 345.3\,\text{kN m}$$
$$M_V = 1053.6\,\text{kN m}$$

The overturning limit state is satisfied, the restoring moment ($M_V$) being greater than the overturning moment ($M_H$). The sliding limit state is satisfied, the resisting force ($V \tan \delta = 227.5\,\text{kN}$) being greater than the disturbing force ($H$).

### 6.5

For $\phi' = 36°$, $K_a = 0.26$ and $K_p = 3.85$.

$$\frac{K_p}{F} = \frac{3.85}{2}$$

$$\gamma' = 20 - 9.8 = 10.2\,\text{kN/m}^3$$

The pressure distribution is shown in Figure Q6.5; hydrostatic pressure on the two sides of the wall balances. Consider moments about X (per m), assuming $d > 0$:

| Force (kN) | | Arm (m) | Moment (kN m) |
|---|---|---|---|
| (1) $\frac{1}{2} \times 0.26 \times 17 \times 4.5^2$ | $= 44.8$ | $d + 1.5$ | $44.8d + 67.2$ |
| (2) $0.26 \times 17 \times 4.5 \times d$ | $= 19.9d$ | $d/2$ | $9.95d^2$ |
| (3) $\frac{1}{2} \times 0.26 \times 10.2 \times d^2$ | $= 1.33d^2$ | $d/3$ | $0.44d^3$ |
| (4) $-\frac{1}{2} \times \frac{3.85}{2} \times 17 \times 1.5^2$ | $= -36.8$ | $d + 0.5$ | $-36.8d - 18.4$ |
| (5) $-\frac{3.85}{2} \times 17 \times 1.5 \times d$ | $= -49.1d$ | $d/2$ | $-24.55d^2$ |
| (6) $-\frac{1}{2} \times \frac{3.85}{2} \times 10.2 \times d^2$ | $= -9.82d^2$ | $d/3$ | $-3.27d^3$ |

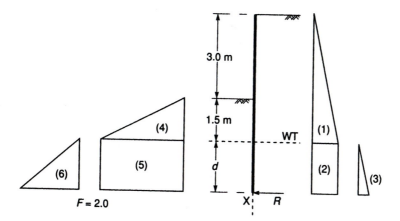

*Figure  Q6.5*

$$\sum M = -2.83d^3 - 14.6d^2 + 8.0d + 48.8 = 0$$
$$\therefore d^3 + 5.16d^2 - 2.83d - 17.24 = 0$$
$$\therefore d = 1.79\,\text{m}$$

Depth of penetration $= 1.2(1.79 + 1.50) = 3.95\,\text{m}$

$$\sum F = 0; \text{ hence } R = 71.5\,\text{kN (substituting } d = 1.79\,\text{m)}$$

Over additional 20% embedded depth:

$$p_p - p_a = (3.85 \times 17 \times 4.5) - (0.26 \times 17 \times 1.5) + (3.85 - 0.26)(10.2 \times 2.12)$$
$$= 365.5\,\text{kN/m}^2$$

Net passive resistance $= 365.5 \times 0.66 = 241\,\text{kN} \,(>R)$

## 6.6

The design value of $\phi' = 36°$, i.e. the partial factor has been applied.
The active pressure coefficient is given by Equation 6.17, in which $\alpha = 105°$, $\beta = 20°$, $\phi = 36°$ and $\delta = 25°$

$$K_a = \left[ \frac{\sin 69°/\sin 105°}{\sqrt{\sin 130°} + \dfrac{\sqrt{(\sin 61° \sin 16°)}}{\sqrt{\sin 85°}}} \right]^2 = 0.50$$

The total active thrust (acting at $25°$ above the normal) is given by Equation 6.16

$$P_a = \frac{1}{2} \times 0.50 \times 19 \times 7.50^2 = 267\,\text{kN/m}$$

Horizontal component:

$$P_h = 267 \cos 40° = 205 \, \text{kN/m}$$

Vertical component:

$$P_v = 267 \sin 40° = 172 \, \text{kN/m}$$

Consider moments about the toe of the wall (Figure Q6.6) (per m):

| Force (kN) | | Arm (m) | Moment (kN m) |
|---|---|---|---|
| (1) $\frac{1}{2} \times 1.75 \times 6.50 \times 23.5$ | $= 133.7$ | 2.58 | 345 |
| (2) $0.50 \times 6.50 \times 23.5$ | $= 76.4$ | 1.75 | 134 |
| (3) $\frac{1}{2} \times 0.70 \times 6.50 \times 23.5$ | $= 53.5$ | 1.27 | 68 |
| (4) $1.00 \times 4.00 \times 23.5$ | $= 94.0$ | 2.00 | 188 |
| (5) $-\frac{1}{2} \times 0.80 \times 0.50 \times 23.5$ | $= -4.7$ | 0.27 | $-1$ |
| $P_a \sin 40°$ | $= 172.0$ $V = \overline{525}$ | 3.33 | $573$ $M_V = \overline{1307}$ |
| $P_a \cos 40°$ | $H = \overline{205}$ | 2.50 | $M_H = 512$ $\Sigma M = \overline{795}$ |

Lever arm of base resultant:

$$\frac{\Sigma M}{V} = \frac{795}{525} = 1.51 \, \text{m}$$

Eccentricity of base resultant:

$$e = 2.00 - 1.51 = 0.49 \, \text{m}$$

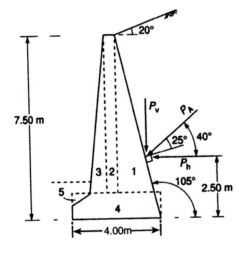

Figure Q6.6

Base pressures (Equation 6.27):

$$p = \frac{525}{4}\left(1 \pm \frac{6 \times 0.49}{4}\right)$$

$$= 228\,\text{kN/m}^2 \text{ and } 35\,\text{kN/m}^2$$

The overturning limit state is satisfied, the restoring moment ($1307\,\text{kN}\,\text{m}$) being greater than the overturning moment ($512\,\text{kN}\,\text{m}$).

The bearing resistance limit state is satisfied, the ultimate bearing capacity of the foundation soil ($250\,\text{kN/m}^2$) being greater than the maximum base pressure ($228\,\text{kN/m}^2$).

The sliding limit state is satisfied, the restoring force ($525 \tan 25° = 245\,\text{kN}$) being greater than the disturbing force ($205\,\text{kN}$).

## 6.7

For $\phi' = 35°$, $K_a = 0.27$;
for $\phi' = 27°$, $K_a = 0.375$ and $K_p = 2.67$;
for soil, $\gamma' = 11.2\,\text{kN/m}^3$;
for backfill, $\gamma' = 10.2\,\text{kN/m}^3$.
The pressure distribution is shown in Figure Q6.7. Hydrostatic pressure is balanced. Consider moments about the anchor point (A), per m:

| Force (kN) | | Arm (m) | Moment (kNm) |
|---|---|---|---|
| (1) $\frac{1}{2} \times 0.27 \times 17 \times 5^2$ | $= 57.4$ | 1.83 | 105.0 |
| (2) $0.27 \times 17 \times 5 \times 3$ | $= 68.9$ | 5.00 | 344.5 |
| (3) $\frac{1}{2} \times 0.27 \times 10.2 \times 3^2$ | $= 12.4$ | 5.50 | 68.2 |
| (4) $0.375\,\{(17 \times 5) + (10.2 \times 3)\}\,d = 43.4d$ | | $d/2 + 6.50$ | $21.7d^2 + 282.1d$ |
| (5) $\frac{1}{2} \times 0.375 \times 11.2 \times d^2$ | $= 2.1d^2$ | $2d/3 + 6.50$ | $1.4d^3 + 13.7d^2$ |
| (6) $-2 \times 10 \times \sqrt{0.375} \times d$ | $= -12.2d$ | $d/2 + 6.50$ | $-6.1d^2 - 79.3d$ |
| (7) $-\frac{1}{2} \times \frac{2.67}{2} \times 11.2 \times d^2$ | $= -7.5d^2$ | $2d/3 + 6.50$ | $-5.0d^3 - 48.8d^2$ |
| (8) $-2 \times 10 \times \frac{\sqrt{2.67}}{2} \times d$ | $= -16.3d$ | $d/2 + 6.50$ | $-8.2d^2 - 106.0d$ |
| Tie rod force per m | $= -T$ | 0 | 0 |

$$\sum M = -3.6d^3 - 27.7d^2 + 96.8d + 517.7 = 0$$

$$\therefore\ d^3 + 7.7d^2 - 26.9d - 143.8 = 0$$

$$\therefore\ d = 4.67\,\text{m}$$

Depth of penetration $= 1.2d = 5.60\,\text{m}$

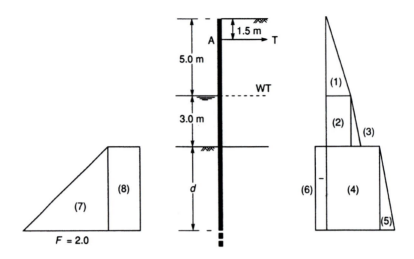

Figure Q6.7

Algebraic sum of forces for $d = 4.67$ m is

$$\sum F = 57.4 + 68.9 + 12.4 + 202.7 + 45.8 - 57.0 - 163.5 - 76.1 - T = 0$$
$$\therefore T = 90.5 \text{ kN/m}$$

Force in each tie rod $= 2.5T = 226$ kN

## 6.8

(a) For $\phi' = 36°$, $K_a = 0.26$ and $K_p = 3.85$;

$$\gamma' = 21 - 9.8 = 11.2 \text{ kN/m}^3$$

The pressure distribution is shown in Figure Q6.8. In this case the net water pressure at C is given by

$$u_C = \frac{15.0}{16.5} \times 1.5 \times 9.8 = 13.4 \text{ kN/m}^2$$

The average seepage pressure is

$$j = \frac{1.5}{16.5} \times 9.8 = 0.9 \text{ kN/m}^3$$

Hence,

$$\gamma' + j = 11.2 + 0.9 = 12.1 \text{ kN/m}^3$$
$$\gamma' - j = 11.2 - 0.9 = 10.3 \text{ kN/m}^3$$

Consider moments about the anchor point A (per m):

| Force (kN) | | | Arm (m) | Moment (kN m) |
|---|---|---|---|---|
| (1) $10 \times 0.26 \times 15.0$ | $= 39.0$ | | 6.0 | 234.0 |
| (2) $\frac{1}{2} \times 0.26 \times 18 \times 4.5^2$ | $= 47.4$ | | 1.5 | 71.1 |
| (3) $0.26 \times 18 \times 4.5 \times 10.5$ | $= 221.1$ | | 8.25 | 1824.0 |
| (4) $\frac{1}{2} \times 0.26 \times 12.1 \times 10.5^2$ | $= 173.4$ | | 10.0 | 1734.0 |
| (5) $\frac{1}{2} \times 13.4 \times 1.5$ | $= 10.1$ | | 4.0 | 40.4 |
| (6) $13.4 \times 3.0$ | $= 40.2$ | | 6.0 | 241.2 |
| (7) $\frac{1}{2} \times 13.4 \times 6.0$ | $= \underline{40.2}$ | | 9.5 | $\underline{381.9}$ |
| | | 571 | | 4527 |
| (8) $-P_{Pm}$ | | | 11.5 | $-11.5 P_{Pm}$ |

$$\sum M = 0$$

$$\therefore P_{Pm} = \frac{4527}{11.5} = 394 \, \text{kN/m}$$

Available passive resistance:

$$P_p = \frac{1}{2} \times 3.85 \times 10.3 \times 6^2 = 714 \, \text{kN/m}$$

Factor of safety:

$$F_p = \frac{P_p}{P_{Pm}} = \frac{714}{394} = 1.8$$

Force in each tie $= 2T = 2(571 - 394) = 354 \, \text{kN}$

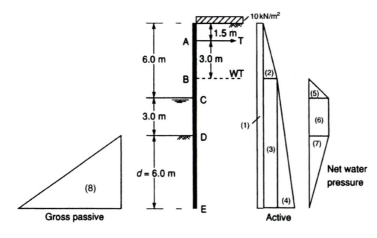

Figure Q6.8

(b) $\phi' = \tan^{-1}(\tan 36°/1.25) = 30°$; therefore, $K_a = 0.33$ and $K_p = 3.0$.
The surcharge is a variable action; therefore, a partial factor of 1.30 applies to force (1).
In this calculation the depth $d$ in Figure Q6.8 is unknown.
  Consider moments (per m) about the tie point A:

| Force (kN) | | Arm (m) |
|---|---|---|
| (1) $0.33 \times 10 \times (d+9.0) \times 1.30 = 4.3d + 38.6$ | | $d/2 + 3.0$ |
| (2) $\frac{1}{2} \times 0.33 \times 18 \times 4.5^2$ $= 60.1$ | | 1.5 |
| (3) $0.33 \times 18 \times 4.5 \times (d+4.5)$ $= 26.7d + 120.3$ | | $d/2 + 5.25$ |
| (4) $\frac{1}{2} \times 0.33 \times 12.1 \times (d+4.5)^2 = 2.00d^2 + 18.0d + 40.4$ | | $2d/3 + 6.0$ |
| (5) $\frac{1}{2} \times 13.4 \times 1.5$ $= 10.1$ | | 4.0 |
| (6) $13.4 \times 3.0$ $= 40.2$ | | 6.0 |
| (7) $\frac{1}{2} \times 13.4 \times d$ $= 6.7d$ | | $d/3 + 7.5$ |
| (8) $-\frac{1}{2} \times 3.0 \times 10.3 \times d^2$ $= -15.45d^2$ | | $2d/3 + 7.5$ |

| Moment (kN m) |
|---|
| (1) $2.15d^2 + 32.2d + 115.8$ |
| (2) 90.2 |
| (3) $13.35d^2 + 200.3d + 631.6$ |
| (4) $1.33d^3 + 24.0d^2 + 134.9d + 242.4$ |
| (5) 40.4 |
| (6) 241.2 |
| (7) $2.20d^2 + 50.2d$ |
| (8) $-10.3d^3 - 115.9d^2$ |

$$\sum M = -8.97d^3 - 74.2d^2 + 417.6d + 1361.6 = 0$$
$$\therefore d^3 + 8.27d^2 - 46.6d - 151.8 = 0$$

By trial,

$$d = 5.44\,\text{m}$$

The minimum depth of embedment required is 5.44 m.


## 6.9

For $\phi' = 30°$ and $\delta = 15°$, $K_a = 0.30$ and $K_p = 4.8$;

$$\gamma' = 20 - 9.8 = 10.2\,\text{kN/m}^3$$

The pressure distribution is shown in Figure Q6.9. Assuming uniform loss in total head along the wall, the net water pressure at C is

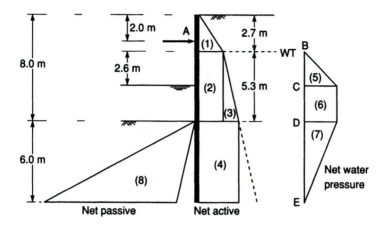

*Figure Q6.9*

$$u_C = \frac{14.7}{17.3} \times 2.6 \times 9.8 = 21.6 \, \text{kN/m}^2$$

and the average seepage pressure around the wall is

$$\bar{j} = \frac{2.6}{17.3} \times 9.8 = 1.5 \, \text{kN/m}^3$$

Consider moments about the prop (A) (per m):

| Force (kN) | | Arm (m) | Moment (kN m) |
|---|---|---|---|
| (1) $\frac{1}{2} \times 0.3 \times 17 \times 2.7^2$ | $= 18.6$ | $-0.20$ | $-3.7$ |
| (2) $0.3 \times 17 \times 2.7 \times 5.3$ | $= 73.0$ | $3.35$ | $244.5$ |
| (3) $\frac{1}{2} \times 0.3 \, (10.2 + 1.5) \, 5.3^2$ | $= 49.3$ | $4.23$ | $208.5$ |
| (4) $0.3 \, \{(17 \times 2.7) + (11.7 \times 5.3)\} \, 6.0$ | $= 194.2$ | $9.00$ | $1747.8$ |
| (5) $\frac{1}{2} \times 21.6 \times 2.6$ | $= 28.1$ | $2.43$ | $68.4$ |
| (6) $21.6 \times 2.7$ | $= 58.3$ | $4.65$ | $271.2$ |
| (7) $\frac{1}{2} \times 21.6 \times 6.0$ | $= 64.8$ | $8.00$ | $518.4$ |
| | | | $\overline{3055}$ |
| (8) $\frac{1}{2} \{4.8 \, (10.2 - 1.5) - 0.3 \, (10.2 + 1.5)\} \, 6.0^2 = 688.5$ | | $10.00$ | $6885$ |

Factor of safety:

$$F_r = \frac{6885}{3055} = 2.25$$

## 6.10

For $\phi' = 40°$, $K_a = 0.22$.
The pressure distribution is shown in Figure Q6.10.

$$p = 0.65K_a\gamma H = 0.65 \times 0.22 \times 19 \times 9 = 24.5\,\text{kN/m}^2$$

Strut load $= 24.5 \times 1.5 \times 3 = 110\,\text{kN}$ (a load factor of at least 2.0 would be applied to this value).
    Using the recommendations of Twine and Roscoe

$$p = 0.2\gamma H = 0.2 \times 19 \times 9 = 34.2\,\text{kN/m}^2$$

Strut load $= 34.2 \times 1.5 \times 3 = 154\,\text{kN}$ (this value would be multiplied by a partial factor of 1.35).

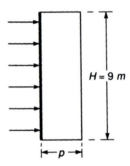

$H = 9\,m$

Figure Q6.10

## 6.11

$$\gamma = 18\,\text{kN/m}^3, \quad \phi' = 34°$$
$$H = 3.50\,\text{m}, \quad nH = 3.35\,\text{m}, \quad mH = 1.85\,\text{m}$$

Consider a trial value of $F = 2.0$. Refer to Figure 6.35.

$$\phi'_m = \tan^{-1}\left(\frac{\tan 34°}{2.0}\right) = 18.6°$$

Then,

$$\alpha = 45° + \frac{\phi'_m}{2} = 54.3°$$

$$W = \frac{1}{2} \times 18 \times 3.50^2 \times \cot 54.3° = 79.2\,\text{kN/m}$$

$$P = \frac{1}{2} \times \gamma_s \times 3.35^2 = 5.61\gamma_s \, \text{kN/m}$$

$$U = \frac{1}{2} \times 9.8 \times 1.85^2 \times \text{cosec } 54.3° = 20.6 \, \text{kN/m}$$

Equations 6.30 and 6.31 then become

$$5.61\gamma_s + (N - 20.6)\tan 18.6° \cos 54.3° - N \sin 54.3° = 0$$
$$79.2 - (N - 20.6)\tan 18.6° \sin 54.3° - N \cos 54.3° = 0$$

i.e.

$$5.61\gamma_s - 0.616N - 4.05 = 0$$
$$79.2 - 0.857N + 5.63 = 0$$
$$\therefore N = \frac{84.8}{0.857} = 98.9 \, \text{kN/m}$$

Then,

$$5.61\gamma_s - 60.9 - 4.05 = 0$$
$$\therefore \gamma_s = \frac{64.9}{5.61} = 11.6 \, \text{kN/m}^3$$

The calculations for trial values of $F$ of 2.0, 1.5 and 1.0 are summarized below:

| $F$ | $\phi'_m$ | $\alpha$ | $W$ (kN/m) | $U$ (kN/m) | $N$ (kN/m) | $\gamma_s$ (kN/m³) |
|-----|-----------|----------|------------|------------|------------|---------------------|
| 2.0 | 18.6° | 54.3° | 79.2 | 20.6 | 98.9 | 11.6 |
| 1.5 | 24.2° | 57.1° | 71.3 | 19.9 | 85.6 | 9.9 |
| 1.0 | 34° | 62° | 58.6 | 19.1 | 65.7 | 7.7 |

$\gamma_s$ is plotted against $F$ in Figure Q6.11.
From Figure Q6.11, for $\gamma_s = 10.6 \, \text{kN/m}^3$, $F = 1.7$

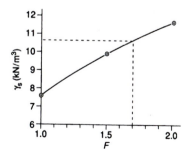

*Figure Q6.11*

## 6.12

For $\phi' = 36°$, $K_a = 0.26$ and $K_0 = 1 - \sin 36° = 0.41$

$$45° + \frac{\phi'}{2} = 63°$$

For the retained material between the surface and a depth of 3.6 m,

$$P_a = \frac{1}{2} \times 0.30 \times 18 \times 3.6^2 = 35.0\,\text{kN/m}$$

Weight of reinforced fill between the surface and a depth of 3.6 m is

$$W_f = 18 \times 3.6 \times 5.0 = 324\,\text{kN/m} \quad = R_v(\text{ or } V)$$

Considering moments about X:

$$(35 \times 1.2) + (324 \times 2.5) = R_v a \quad (a = \text{lever arm})$$
$$\therefore a = 2.63\,\text{m}$$

Eccentricity of $R_v$:

$$e = 2.63 - 2.50 = 0.13\,\text{m}$$

The average vertical stress at a depth of 3.6 m is

$$\sigma_z = \frac{R_v}{L - 2e} = \frac{324}{4.74} = 68\,\text{kN/m}^2$$

(a) In the tie back wedge method, $K = K_a$ and $L_e = 4.18\,\text{m}$

$$\therefore T_p = 0.26 \times 68 \times 1.20 \times 0.65 = 13.8\,\text{kN} \quad \text{(Equation 6.32)}$$

$$\text{Tensile stress in the element} = \frac{13.8 \times 10^3}{65 \times 3} = 71\,\text{N/mm}^2$$

Tensile failure of the element will not occur, the ultimate tensile strength being 4.8 times the tensile stress.

$$T_r = 2 \times 0.065 \times 4.18 \times 68 \times \tan 30° = 21.3\,\text{kN}$$

Slipping between element and soil will not occur, $T_r$ being greater than $T_p$ by a factor of 1.5. The stability of wedge ABC would also be checked after the above calculations had been performed for all elements.
    The weight of ABC is

$$W = \frac{1}{2} \times 18 \times 5.2 \times 2.65 = 124\,\text{kN/m}$$

From the force diagram (Figure Q6.12), $T_w = 124 \tan 27° = 63.2\,\text{kN}$
    The factored sum of the forces $T_r$ in all elements $(\Sigma T_r)$ must be greater than $T_w$.

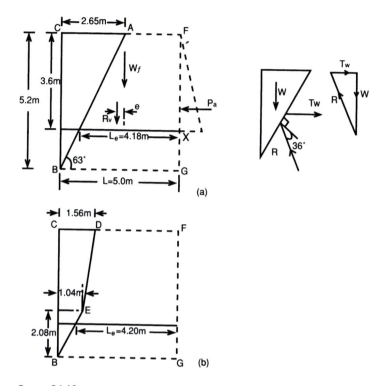

Figure Q6.12

(b) In the coherent gravity method, $K = 0.32$ ($K$ varying linearly between 0.41 at the surface and 0.26 at a depth of 6 m) and $L_e = 4.20$ m.

$$\therefore \quad T_p = 0.32 \times 68 \times 1.20 \times 0.65 = 17.0 \,\text{kN}$$

$$T_r = 21.3 \times \frac{4.20}{4.18} = 21.4 \,\text{kN}$$

Again, the tensile failure and slipping limit states are satisfied for this element.

# Chapter 7

# Consolidation theory

## 7.1

Total change in thickness:

$$\Delta H = 7.82 - 6.02 = 1.80 \, \text{mm}$$

$$\text{Average thickness} = 15.30 + \frac{1.80}{2} = 16.20 \, \text{mm}$$

$$\text{Length of drainage path, } d = \frac{16.20}{2} = 8.10 \, \text{mm}$$

Root time plot (Figure Q7.1a):

$$\sqrt{t_{90}} = 3.3$$
$$\therefore t_{90} = 10.9 \, \text{min}$$

$$c_v = \frac{0.848 d^2}{t_{90}} = \frac{0.848 \times 8.10^2}{10.9} \times \frac{1440 \times 365}{10^6} = 2.7 \, \text{m}^2/\text{year}$$

$$r_0 = \frac{7.82 - 7.64}{7.82 - 6.02} = \frac{0.18}{1.80} = 0.100$$

$$r_p = \frac{10(7.64 - 6.45)}{9(7.82 - 6.02)} = \frac{10 \times 1.19}{9 \times 1.80} = 0.735$$

$$r_s = 1 - (0.100 + 0.735) = 0.165$$

Log time plot (Figure Q7.1b):

$$t_{50} = 2.6 \, \text{min}$$

$$c_v = \frac{0.196 d^2}{t_{50}} = \frac{0.196 \times 8.10^2}{2.6} \times \frac{1440 \times 365}{10^6} = 2.6 \, \text{m}^2/\text{year}$$

$$r_0 = \frac{7.82 - 7.63}{7.82 - 6.02} = \frac{0.19}{1.80} = 0.106$$

$$r_p = \frac{7.63 - 6.23}{7.82 - 6.02} = \frac{1.40}{1.80} = 0.778$$

$$r_s = 1 - (0.106 + 0.778) = 0.116$$

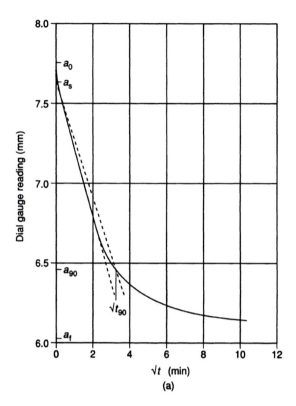

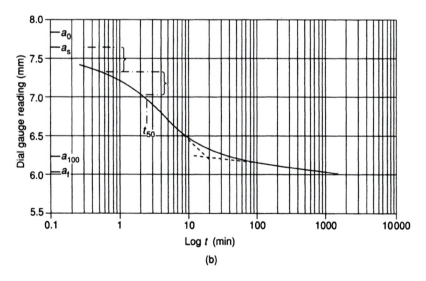

Figure Q7.1(a)

Figure Q7.1(b)

Final void ratio:

$$e_1 = w_1 G_s = 0.232 \times 2.72 = 0.631$$

$$\frac{\Delta e}{\Delta H} = \frac{1 + e_0}{H_0} = \frac{1 + e_1 + \Delta e}{H_0}$$

i.e.

$$\frac{\Delta e}{1.80} = \frac{1.631 + \Delta e}{17.10}$$

$$\therefore \Delta e = \frac{2.936}{15.30} = 0.192$$

Initial void ratio, $e_0 = 0.631 + 0.192 = 0.823$
   Then,

$$m_v = \frac{1}{1 + e_0} \times \frac{e_0 - e_1}{\sigma_1' - \sigma_0'} = \frac{1}{1.823} \times \frac{0.192}{0.107} = 0.98\,\mathrm{m^2/MN}$$

$$k = c_v m_v \gamma_w = \frac{2.65 \times 0.98 \times 9.8}{60 \times 1440 \times 365 \times 10^3} = 8.1 \times 10^{-10}\,\mathrm{m/s}$$

## 7.2

Using Equation 7.7 (one-dimensional method):

$$S_c = \frac{e_0 - e_1}{1 + e_0} H$$

Appropriate values of $e$ are obtained from Figure Q7.2. The clay will be divided into four sublayers, hence $H = 2000\,\mathrm{mm}$.

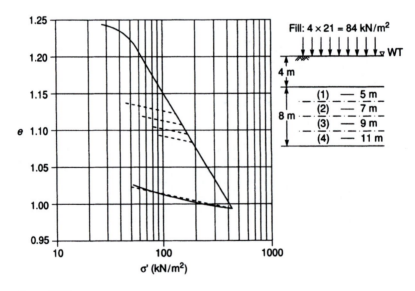

Figure Q7.2

*Settlement*

| Layer | $\sigma_0'$ (kN/m$^2$) | $\sigma_1'$ (kN/m$^2$) | $e_0$ | $e_1$ | $e_0 - e_1$ | $s_c$ (mm) |
|---|---|---|---|---|---|---|
| 1 | 46.0* | 130.0$^\dagger$ | 1.236 | 1.123 | 0.113 | 101 |
| 2 | 64.4 | 148.4 | 1.200 | 1.108 | 0.092 | 84 |
| 3 | 82.8 | 166.8 | 1.172 | 1.095 | 0.077 | 71 |
| 4 | 101.2 | 185.2 | 1.150 | 1.083 | 0.067 | 62 |
| | | | | | | 318 |

Notes
* 5 × 9.2.
† 46.0 + 84.

*Heave*

| Layer | $\sigma_0'$ (kN/m$^2$) | $\sigma_1'$ (kN/m$^2$) | $e_0$ | $e_1$ | $e_0 - e_1$ | $s_c$ (mm) |
|---|---|---|---|---|---|---|
| 1 | 130.0 | 46.0 | 1.123 | 1.136 | −0.013 | −12 |
| 2 | 148.4 | 64.4 | 1.108 | 1.119 | −0.011 | −10 |
| 3 | 166.8 | 82.8 | 1.095 | 1.104 | −0.009 | −9 |
| 4 | 185.2 | 101.2 | 1.083 | 1.091 | −0.008 | −7 |
| | | | | | | −38 |

**7.3**

$$U = f(T_v) = f\left(\frac{c_v t}{d^2}\right)$$

Hence if $c_v$ is constant,

$$\frac{t_1}{t_2} = \frac{d_1^2}{d_2^2}$$

where '1' refers to the oedometer specimen and '2' the clay layer.
   For open layers:

$$d_1 = 9.5\,\text{mm} \quad \text{and} \quad d_2 = 2500\,\text{mm}$$

$\therefore$ for $U = 0.50$, $t_2 = t_1 \times \dfrac{d_2^2}{d_1^2}$

$$= \frac{20}{60 \times 24 \times 365} \times \frac{2500^2}{9.5^2} = 2.63\,\text{years}$$

for $U < 0.60$, $T_v = \dfrac{\pi}{4} U^2$   (Equation 7.24(a))

$\therefore t_{0.30} = t_{0.50} \times \dfrac{0.30^2}{0.50^2}$

$$= 2.63 \times 0.36 = 0.95\,\text{years}$$

## 7.4

The layer is open,

$$\therefore d = \frac{8}{2} = 4\,\text{m}$$

$$T_v = \frac{c_v t}{d^2} = \frac{2.4 \times 3}{4^2} = 0.450$$

$$u_i = \Delta\sigma = 84\,\text{kN/m}^2$$

The excess pore water pressure is given by Equation 7.21:

$$u_e = \sum_{m=0}^{m=\infty} \frac{2u_i}{M} \left[ \sin\left(\frac{Mz}{d}\right) \right] \exp(-M^2 T_v)$$

In this case, $z = d$:

$$\therefore \sin\left(\frac{Mz}{d}\right) = \sin M$$

where

$$M = \frac{\pi}{2}, \frac{3\pi}{2}, \frac{5\pi}{2}, \dots$$

| $M$ | sin $M$ | $M^2 T_v$ | exp $(-M^2 T_v)$ |
|---|---|---|---|
| $\dfrac{\pi}{2}$ | $+1$ | 1.110 | 0.329 |
| $\dfrac{3\pi}{2}$ | $-1$ | 9.993 | $4.57 \times 10^{-5}$ |

$$\therefore u_e = 2 \times 84 \times \frac{2}{\pi} \times 1 \times 0.329 \quad \text{(other terms negligible)}$$

$$= 35.2\,\text{kN/m}^2$$

## 7.5

The layer is open,

$$\therefore d = \frac{6}{2} = 3\,\text{m}$$

$$T_v = \frac{c_v t}{d^2} = \frac{1.0 \times 3}{3^2} = 0.333$$

The layer thickness will be divided into six equal parts, i.e. $m = 6$.

For an open layer:

$$T_v = 4\frac{n}{m^2}\beta$$

$$\therefore n\beta = \frac{0.333 \times 6^2}{4} = 3.00$$

The value of $n$ will be taken as 12 (i.e. $\Delta t = 3/12 = 1/4$ year), making $\beta = 0.25$. The computation is set out below, all pressures having been multiplied by 10:

$$u_{i,j+1} = u_{i,j} + 0.25(u_{i-1,j} + u_{i+1,j} - 2u_{i,j})$$

| $i$ | $j$ | | | | | | | | | | | | |
|---|---|---|---|---|---|---|---|---|---|---|---|---|---|
| | 0 | 1 | 2 | 3 | 4 | 5 | 6 | 7 | 8 | 9 | 10 | 11 | 12 |
| 0 | 0 | 0 | 0 | 0 | 0 | 0 | 0 | 0 | 0 | 0 | 0 | 0 | 0 |
| 1 | 500 | 350 | 275 | 228 | 195 | 171 | 151 | 136 | 123 | 112 | 102 | 94 | 87 |
| 2 | 400 | 400 | 362 | 325 | 292 | 264 | 240 | 219 | 201 | 185 | 171 | 158 | 146 |
| 3 | 300 | 300 | 300 | 292 | 277 | 261 | 245 | 230 | 215 | 201 | 187 | 175 | 163 |
| 4 | 200 | 200 | 200 | 200 | 198 | 193 | 189 | 180 | 171 | 163 | 154 | 145 | 137 |
| 5 | 100 | 100 | 100 | 100 | 100 | 99.5 | 98 | 96 | 93 | 89 | 85 | 81 | 77 |
| 6 | 0 | 0 | 0 | 0 | 0 | 0 | 0 | 0 | 0 | 0 | 0 | 0 | 0 |

The initial and 3-year isochrones are plotted in Figure Q7.5.

Area under initial isochrone = 180 units

Area under 3-year isochrone = 63 units

The average degree of consolidation is given by Equation 7.25. Thus

$$U = 1 - \frac{63}{180} = 0.65$$

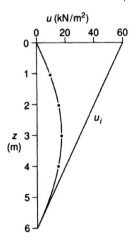

Figure Q7.5

## 7.6

At the top of the clay layer the decrease in pore water pressure is $4\gamma_w$. At the bottom of the clay layer the pore water pressure remains constant. Hence at the centre of the clay layer,

$$\Delta\sigma' = 2\gamma_w = 2 \times 9.8 = 19.6 \, \text{kN/m}^2$$

The final consolidation settlement (one-dimensional method) is

$$s_c = m_v \Delta\sigma' H = 0.83 \times 19.6 \times 8 = 130 \, \text{mm}$$

$$\text{Corrected time, } t = 2 - \frac{1}{2}\left(\frac{40}{52}\right) = 1.615 \, \text{years}$$

$$\therefore \, T_v = \frac{c_v t}{d^2} = \frac{4.4 \times 1.615}{4^2} = 0.444$$

From Figure 7.18 (curve 1), $U = 0.73$.
Settlement after 2 years $= U s_c = 0.73 \times 130 = 95 \, \text{mm}$

## 7.7

The clay layer is thin relative to the dimensions of the raft, and therefore the one-dimensional method is appropriate. The clay layer can be considered as a whole (see Figure Q7.7)

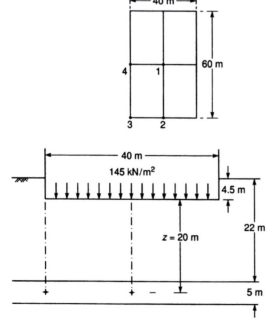

Figure Q7.7

| Point | $m$ | $n$ | $I_r$ | $\Delta\sigma$ (kN/m²) | $s_c^*$ (mm) |
|---|---|---|---|---|---|
| 1 | $\dfrac{30}{20} = 1.5$ | $\dfrac{20}{20} = 1.0$ | 0.194 (×4) | 113 | 124 |
| 2 | $\dfrac{60}{20} = 3.0$ | $\dfrac{20}{20} = 1.0$ | 0.204 (×2) | 59 | 65 |
| 3 | $\dfrac{60}{20} = 3.0$ | $\dfrac{40}{20} = 2.0$ | 0.238 (×1) | 35 | 38 |
| 4 | $\dfrac{30}{20} = 1.5$ | $\dfrac{40}{20} = 2.0$ | 0.224 (×2) | 65 | 72 |

Note
* $s_c = m_v \Delta\sigma' H = 0.22 \times \Delta\sigma' \times 5 = 1.1\Delta\sigma'$ (mm)  $(\Delta\sigma' = \Delta\sigma)$.

## 7.8

Due to the thickness of the clay layer relative to the size of the foundation, there will be significant lateral strain in the clay and the Skempton–Bjerrum method is appropriate. The clay is divided into six sublayers (Figure Q7.8) for the calculation of consolidation settlement.

(a) Immediate settlement:

$$\frac{H}{B} = \frac{30}{35} = 0.86$$
$$\frac{D}{B} = \frac{2}{35} = 0.06$$

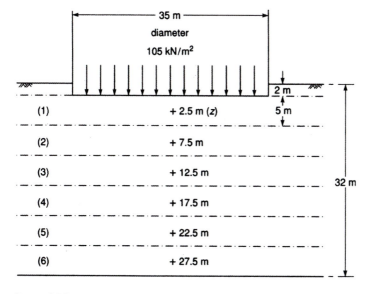

Figure Q7.8

From Figure 5.15 (circle), $\mu_1 = 0.32$ and $\mu_0 = 1.0$:

$$s_i = \mu_0 \mu_1 \frac{qB}{E_u} = 1.0 \times 0.32 \times \frac{105 \times 35}{40} = 30\,\text{mm}$$

(b) Consolidation settlement:

| Layer | $z$ (m) | $D/z$ | $I_c^*$ | $\Delta\sigma$ (kN/m²) | $s_{od}^\dagger$ (mm) |
|---|---|---|---|---|---|
| I | 2.5 | 14 | 0.997 | 107 | 73.5 |
| 2 | 7.5 | 4.67 | 0.930 | 98 | 68.6 |
| 3 | 12.5 | 2.80 | 0.804 | 84 | 58.8 |
| 4 | 17.5 | 2.00 | 0.647 | 68 | 47.6 |
| 5 | 22.5 | 1.55 | 0.505 | 53 | 37.1 |
| 6 | 27.5 | 1.27 | 0.396 | 42 | 29.4 |
| | | | | | 315.0 |

Notes
* From Figure 5.9.
† $s_{od} = m_v \Delta\sigma' H = 0.14 \times \Delta\sigma' \times 5 = 0.70\Delta\sigma'$   $(\Delta\sigma' = \Delta\sigma)$.

Now

$$\frac{H}{B} = \frac{30}{35} = 0.86 \quad \text{and} \quad A = 0.65$$

$\therefore$ from Figure 7.12, $\mu = 0.79$

$\therefore s_c = \mu s_{od} = 0.79 \times 315 = 250\,\text{mm}$

Total settlement:

$$s = s_i + s_c$$
$$= 30 + 250 = 280\,\text{mm}$$

## 7.9

Without sand drains:

$$U_v = 0.25$$

$\therefore T_v = 0.049$ (from Figure 7.18)

$$\therefore t = \frac{T_v d^2}{c_v} = \frac{0.049 \times 8^2}{c_v}$$

With sand drains:

$$R = 0.564S = 0.564 \times 3 = 1.69\,\text{m}$$

$$n = \frac{R}{r} = \frac{1.69}{0.15} = 11.3$$

$$T_r = \frac{c_h t}{4R^2} = \frac{c_h}{4 \times 1.69^2} \times \frac{0.049 \times 8^2}{c_v} \quad (\text{and } c_h = c_v)$$

$$= 0.275$$

$\therefore U_r = 0.73$   (from Figure 7.30)

Using Equation 7.40:

$$(1 - U) = (1 - U_v)(1 - U_r)$$
$$= (1 - 0.25)(1 - 0.73) = 0.20$$
$$\therefore U = 0.80$$

## 7.10

Without sand drains:

$$U_v = 0.90$$
$$\therefore T_v = 0.848$$
$$\therefore t = \frac{T_v d^2}{c_v} = \frac{0.848 \times 10^2}{9.6} = 8.8 \text{ years}$$

With sand drains:

$$R = 0.564S = 0.564 \times 4 = 2.26\,\text{m}$$
$$n = \frac{R}{r} = \frac{2.26}{0.15} = 15$$
$$\frac{T_r}{T_v} = \frac{c_h}{c_v} \frac{d^2}{4R^2} \quad (\text{same } t)$$
$$\therefore \frac{T_r}{T_v} = \frac{14.0}{9.6} \times \frac{10^2}{4 \times 2.26^2} = 7.14 \tag{1}$$

Using Equation 7.40:

$$(1 - U) = (1 - U_v)(1 - U_r)$$
$$\therefore (1 - 0.90) = (1 - U_v)(1 - U_r)$$
$$\therefore (1 - U_v)(1 - U_r) = 0.10 \tag{2}$$

An iterative solution is required using (1) and (2), an initial value of $U_v$ being estimated.

| $U_v$ | $T_v$ | $T_r = 7.14T_v$ | $U_r$ | $(1 - U_v)(1 - U_r)$ |
|---|---|---|---|---|
| 0.40 | 0.1256 | 0.897 | 0.97 | $0.60 \times 0.03 = 0.018$ |
| 0.30 | 0.0707 | 0.505 | 0.87 | $0.70 \times 0.13 = 0.091$ |
| 0.29 | 0.0660 | 0.471 | 0.85 | $0.71 \times 0.15 = 0.107$ |
| 0.295 | 0.0683 | 0.488 | 0.86 | $0.705 \times 0.14 = 0.099$ |

Thus

$$U_v = 0.295 \text{ and } U_r = 0.86$$
$$\therefore t = 8.8 \times \frac{0.0683}{0.848} = 0.7 \text{ years}$$

# Chapter 8

# Bearing capacity

## 8.1

(a) The ultimate bearing capacity is given by Equation 8.3

$$q_f = cN_c + \gamma DN_q + \frac{1}{2}\gamma BN_\gamma$$

For $\phi_u = 0$:

$$N_c = 5.14, \ N_q = 1, \ N_\gamma = 0$$
$$\therefore q_f = (105 \times 5.14) + (21 \times 1 \times 1) = 540 + 21 \text{ kN/m}^2$$

The net ultimate bearing capacity is

$$q_{nf} = q_f - \gamma D = 540 \text{ kN/m}^2$$

The net foundation pressure is

$$q_n = q - \gamma D = \frac{425}{2} - (21 \times 1) = 192 \text{ kN/m}^2$$

The factor of safety (Equation 8.6) is

$$F = \frac{q_{nf}}{q_n} = \frac{540}{192} = 2.8$$

(b) For $\phi' = 28°$:

$$N_c = 26, \ N_q = 15, \ N_\gamma = 13 \qquad \text{(from Figure 8.4)}$$
$$\gamma' = 21 - 9.8 = 11.2 \text{ kN/m}^3$$
$$\therefore q_f = (10 \times 26) + (11.2 \times 1 \times 15) + \left(\frac{1}{2} \times 11.2 \times 2 \times 13\right)$$
$$= 260 + 168 + 146 = 574 \text{ kN/m}^2$$

$q_{nf} = 574 - 11.2 = 563 \, \text{kN/m}^2$

$F = \dfrac{563}{192} = 2.9$

($q_n = 192 \, \text{kN/m}^2$ assumes that backfilled soil on the footing slab is included in the load of 425 kN/m.)

## 8.2

For $\phi' = 38°$:

$N_q = 49, \quad N_\gamma = 67$

$\therefore q_{nf} = \gamma D(N_q - 1) + \dfrac{1}{2}\gamma B N_\gamma \qquad \text{(from Equation 8.3)}$

$\quad = (18 \times 0.75 \times 48) + \left(\dfrac{1}{2} \times 18 \times 1.5 \times 67\right)$

$\quad = 648 + 905 = 1553 \, \text{kN/m}^2$

$q_n = \dfrac{500}{1.5} - (18 \times 0.75) = 320 \, \text{kN/m}^2$

$\therefore F = \dfrac{q_{nf}}{q_n} = \dfrac{1553}{320} = 4.8$

$\phi'_d = \tan^{-1}\left(\dfrac{\tan 38°}{1.25}\right) = 32°$, therefore $N_q = 23$ and $N_\gamma = 25$.

Design bearing resistance, $R_d = 1.5\left[(18 \times 0.75 \times 23) + \left(\dfrac{1}{2} \times 18 \times 1.5 \times 25\right)\right]$

$\quad = 1.5(310 + 337)$

$\quad = 970 \, \text{kN/m}$

Design load (action), $V_d = 500 \, \text{kN/m}$

The design bearing resistance is greater than the design load, therefore the bearing resistance limit state is satisfied.

## 8.3

$\dfrac{D}{B} = \dfrac{3.50}{2.25} = 1.55$

From Figure 8.5, for a square foundation:

$N_c = 8.1$

For a rectangular foundation ($L = 4.50$ m; $B = 2.25$ m):

$$N_c = \left(0.84 + 0.16\frac{B}{L}\right)8.1 = 7.45$$

Using Equation 8.10:

$$q_{nf} = q_f - \gamma D = c_u N_c = 135 \times 7.45 = 1006 \, \text{kN/m}^2$$

For $F = 3$:

$$q_n = \frac{1006}{3} = 335 \, \text{kN/m}^2$$

$$\therefore q = q_n + \gamma D = 335 + (20 \times 3.50) = 405 \, \text{kN/m}^2$$

$$\therefore \text{Design load} = 405 \times 4.50 \times 2.25 = 4100 \, \text{kN}$$

$$\text{Design undrained strength, } c_{ud} = \frac{135}{1.4} = 96 \, \text{kN/m}^2$$

$$\text{Design bearing resistance, } R_d = c_{ud} N_c \times \text{area} = 96 \times 7.45 \times 4.50 \times 2.25$$
$$= 7241 \, \text{kN}$$

$$\text{Design load, } V_d = 4100 \, \text{kN}$$

$R_d > V_d$, therefore the bearing resistance limit state is satisfied.

## 8.4

For $\phi' = 40°$:

$$N_q = 64, \ N_\gamma = 95$$
$$q_{nf} = \gamma D(N_q - 1) + 0.4\gamma B N_\gamma$$

(a) Water table 5 m below ground level:

$$q_{nf} = (17 \times 1 \times 63) + (0.4 \times 17 \times 2.5 \times 95)$$
$$= 1071 + 1615 = 2686 \, \text{kN/m}^2$$
$$q_n = 400 - 17 = 383 \, \text{kN/m}^2$$
$$F = \frac{2686}{383} = 7.0$$

(b) Water table 1 m below ground level (i.e. at foundation level):

$$\gamma' = 20 - 9.8 = 10.2 \, \text{kN/m}^3$$

$$q_{nf} = (17 \times 1 \times 63) + (0.4 \times 10.2 \times 2.5 \times 95)$$
$$= 1071 + 969 = 2040\,kN/m^2$$
$$F = \frac{2040}{383} = 5.3$$

(c) Water table at ground level with upward hydraulic gradient 0.2:

$$(\gamma' - j) = 10.2 - (0.2 \times 9.8) = 8.2\,kN/m^3$$
$$q_{nf} = (8.2 \times 1 \times 63) + (0.4 \times 8.2 \times 2.5 \times 95)$$
$$= 517 + 779 = 1296\,kN/m^2$$
$$F = \frac{1296}{392} = 3.3$$

## 8.5

The following partial factors are used: dead load, 1.0; imposed load, 1.3; shear strength ($\tan \phi'$), 1.25.

Design load, $V_d = 4000 + (1.3 \times 1500) = 5950\,kN$

Design value of $\phi' = \tan^{-1}\left(\dfrac{\tan 39°}{1.25}\right) = 33°$

For $\phi' = 33°$, $N_q = 26$ and $N_\gamma = 29$:

$$\text{Design bearing resistance,}\ R_d = 3^2[(10.2 \times 1.5 \times 26) + (0.4 \times 10.2 \times 3.0 \times 29)]$$
$$= 3^2(398 + 355)$$
$$= 6777\,kN$$

$R_d > V_d$, therefore the bearing resistance limit state is satisfied.

## 8.6

(a) Undrained shear, for $\phi_u = 0$:

$$N_c = 5.14,\ N_q = 1,\ N_\gamma = 0$$
$$q_{nf} = 1.2c_u N_c$$
$$= 1.2 \times 100 \times 5.14 = 617\,kN/m^2$$
$$q_n = \frac{q_{nf}}{F} = \frac{617}{3} = 206\,kN/m^2$$
$$q = q_n + \gamma D = 206 + 21 = 227\,kN/m^2$$

Drained shear, for $\phi' = 32°$:

$N_q = 23, \ N_\gamma = 25$

$\gamma' = 21 - 9.8 = 11.2 \, \text{kN/m}^3$

$q_{nf} = \gamma' D(N_q - 1) + 0.4\gamma' B N_\gamma$

$\quad = (11.2 \times 1 \times 22) + (0.4 \times 11.2 \times 4 \times 25)$

$\quad = 246 + 448$

$\quad = 694 \, \text{kN/m}^2$

$q = \dfrac{694}{3} + 21 = 231 + 21 = 252 \, \text{kN/m}^2$

Design load $= 4^2 \times 227 = 3632 \, \text{kN}$

(b) Design undrained strength, $c_{ud} = \dfrac{100}{1.4} = 71 \, \text{kN/m}^2$

$\quad$ Design bearing resistance, $R_d = 1.2 c_{ud} N_e \times \text{area}$

$\qquad\qquad\qquad\qquad\qquad\quad = 1.2 \times 71 \times 5.14 \times 4^2$

$\qquad\qquad\qquad\qquad\qquad\quad = 7007 \, \text{kN}$

For drained shear, $\phi'_d = \tan^{-1}\left(\dfrac{\tan 32°}{1.25}\right) = 26°$

$\therefore \ N_q = 12, \ N_\gamma = 10$

$\quad$ Design bearing resistance, $R_d = 4^2[(11.2 \times 1 \times 12) + (0.4 \times 11.2 \times 4 \times 10)]$

$\qquad\qquad\qquad\qquad\qquad\qquad\quad = 4^2(134 + 179)$

$\qquad\qquad\qquad\qquad\qquad\qquad\quad = 5008 \, \text{kN}$

(c) Consolidation settlement: the clay will be divided into three sublayers (Figure Q8.6):

| Layer | z (m) | m, n | $I_r$ | $\Delta\sigma'$ (kN/m²) | $s_{od}$ (mm) |
|-------|-------|------|-------|------------------------|---------------|
| 1 | 2 | 1.00 | 0.175 | $0.700q_n$ | $0.182q_n$ |
| 2 | 6 | 0.33 | 0.044 | $0.176q_n$ | $0.046q_n$ |
| 3 | 10 | 0.20 | 0.017 | $0.068q_n$ | $0.018q_n$ |
|   |   |   |   |   | $0.246q_n$ |

Diameter of equivalent circle, $B = 4.5 \, \text{m}$

$\dfrac{H}{B} = \dfrac{12}{4.5} = 2.7 \quad \text{and} \quad A = 0.42$

$\therefore \ \mu = 0.60 \quad \text{(from Figure 7.12)}$

$s_c = 0.60 \times 0.246q_n = 0.147q_n \quad \text{(mm)}$

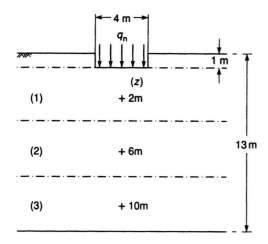

*Figure Q8.6*

For $s_c = 30$ mm:

$$q_n = \frac{30}{0.147} = 204\,\text{kN/m}^2$$

$$q = 204 + 21 = 225\,\text{kN/m}^2$$

$$\text{Design load} = 4^2 \times 225 = 3600\,\text{kN}$$

The design load is 3600 kN, settlement being the limiting criterion.

**8.7**

$$\frac{D}{B} = \frac{8}{4} = 2.0$$

From Figure 8.5, for a strip, $N_c = 7.1$. For a depth/breadth ratio of 2, Equation 8.12 should be used.

$$F = \frac{c_u N_c}{\gamma D} = \frac{40 \times 7.1}{20 \times 8} = 1.8$$

**8.8**

Design load for ultimate limit state, $V_d = 2500 + (1250 \times 1.30) = 4125\,\text{kN}$

Design value of $\phi' = \tan^{-1}\left(\dfrac{\tan 38°}{1.25}\right) = 32°$

For $\phi' = 32°$, $N_q = 23$ and $N_\gamma = 25$

$$\text{Design bearing resistance, } R_d = 2.50^2[(17 \times 1.0 \times 23) + (0.4 \times 10.2 \times 2.50 \times 25)]$$
$$= 2.50^2(391 + 255)$$
$$= 4037\,\text{kN}$$

The design bearing resistance is (slightly) less than the design load, therefore the bearing resistance limit state is not satisfied. To satisfy the limit state, the dimension of the foundation should be increased to 2.53 m.

Design load for serviceability limit state $= 2500 + 1250 = 3750\,\text{kN}$

For $B = 2.50\,\text{m}$, $q_n = \dfrac{3750}{2.50^2} - 17 = 583\,\text{kN/m}^2$

From Figure 5.10, $m = n = \dfrac{1.26}{6} = 0.21$

$\therefore I_r = 0.019$

Stress increment, $\Delta\sigma = 4 \times 0.019 \times 583 = 44\,\text{kN/m}^2$

Consolidation settlement, $s_c = m_v\Delta\sigma H = 0.15 \times 44 \times 2 = 13\,\text{mm}$

The settlement would be less than 13 mm if an appropriate value of settlement coefficient (Figure 7.12) was applied.
    The settlement is less than 20 mm, therefore the serviceability limit state is satisfied.

## 8.9

| Depth (m) | N | $\sigma'_v$ (kN/m²)* | $C_N$ | $N_1$ |
|---|---|---|---|---|
| 0.70 | 6 | – | – | – |
| 1.35 | 9 | 23 | 1.90 | 17 |
| 2.20 | 10 | 37 | 1.55 | 15 |
| 2.95 | 8 | 50 | 1.37 | 11 |
| 3.65 | 12 | 58 | 1.28 | 15 |
| 4.40 | 13 | 65 | 1.23 | 16 |
| 5.15 | 17 | – | – | – |
| 6.00 | 23 | – | – | – |

Note
* Using $\gamma = 17\,\text{kN/m}^3$ and $\gamma' = 10\,\text{kN/m}^3$.

(a) *Terzaghi and Peck.* Use $N_1$ values between depths of 1.2 and 4.7 m, the average value being 15. For $B = 3.5\,\text{m}$ and $N = 15$, the provisional value of bearing capacity, using Figure 8.10, is $150\,\text{kN/m}^2$. The water table correction factor (Equation 8.16) is

$$C_w = 0.5 + 0.5\left(\frac{3.0}{4.7}\right) = 0.82$$

Thus

$$q_a = 150 \times 0.82 = 120\,\text{kN/m}^2$$

(b) *Meyerhof.* Use uncorrected $N$ values between depths of 1.2 and 4.7 m, the average value being 10. For $B = 3.5$ m and $N = 10$, the provisional value of bearing capacity, using Figure 8.10, is 90 kN/m². This value is increased by 50%.
   Thus

$$q_a = 90 \times 1.5 = 135\,\text{kN/m}^2$$

(c) *Burland and Burbidge.* Using Figure 8.12, for $B = 3.5$ m, $z_1 = 2.5$ m. Use $N$ values between depths of 1.2 and 3.7 m, the average value being 10. From Equation 8.18:

$$I_c = \frac{1.71}{10^{1.4}} = 0.068$$

From Equation 8.19(a), with $s = 25$ mm:

$$q = \frac{25}{3.5^{0.7} \times 0.068} = 150\,\text{kN/m}^2$$

## 8.10

Peak value of strain influence factor occurs at a depth of 2.7 m and is given by

$$I_{zp} = 0.5 + 0.1\left(\frac{130}{16 \times 2.7}\right)^{0.5} = 0.67$$

Refer to Figure Q8.10:

$$E = 2.5 q_c$$

| Layer | $\Delta z$ (m) | $q_c$ (MN/m²) | $E = 2.5 q_c$ (MN/m²) | $I_z$ | $\frac{I_z}{E}\Delta z$ (mm³/MN) |
|---|---|---|---|---|---|
| 1 | 1.2 | 2.6 | 6.5 | 0.33 | 0.061 |
| 2 | 0.8 | 5.0 | 12.5 | 0.65 | 0.042 |
| 3 | 1.6 | 4.0 | 10.0 | 0.48 | 0.077 |
| 4 | 1.6 | 7.2 | 18.0 | 0.24 | 0.021 |
| 5 | 0.8 | 12.4 | 31.0 | 0.07 | 0.002 |
| | | | | | 0.203 |

$$C_1 = 1 - 0.5\frac{\sigma_0'}{q_n} = 1 - \frac{0.5 \times 1.2 \times 16}{130} = 0.93$$

$$C_2 = 1 \quad (\text{say})$$

$$\therefore\ s = C_1 C_2 q_n \sum \frac{I_z}{E}\Delta z = 0.93 \times 1 \times 130 \times 0.203 = 25\,\text{mm}$$

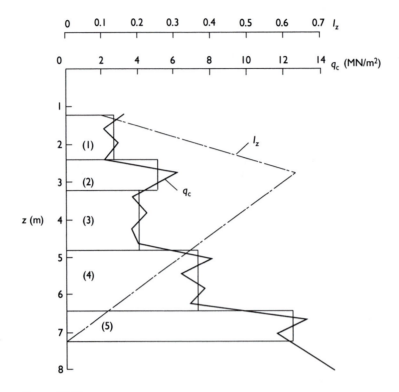

Figure Q8.10

## 8.11

At pile base level:

$$c_u = 220 \, \text{kN/m}^2$$
$$\therefore q_b = c_u N_c = 220 \times 9 = 1980 \, \text{kN/m}^2$$

Disregard skin friction over a length of $2B$ above the under-ream. Between 4 and 17.9 m:

$$\overline{\sigma}_0' = 10.95 \gamma' = 10.95 \times 11.2 = 122.6 \, \text{kN/m}^2$$
$$\therefore q_s = \beta \overline{\sigma}_0' = 0.7 \times 122.6 = 86 \, \text{kN/m}^2$$

Then

$$Q_f = A_b q_b + A_s q_s$$
$$= \left( \frac{\pi}{4} \times 3^2 \times 1980 \right) + (\pi \times 1.05 \times 13.9 \times 86)$$
$$= 13\,996 + 3941 = 17\,937 \, \text{kN}$$

Allowable load:

(a)  $\dfrac{Q_f}{2} = \dfrac{17\,937}{2} = 8968\,\text{kN}$

(b)  $\dfrac{A_b q_b}{3} + A_s q_s = \dfrac{13\,996}{3} + 3941 = 8606\,\text{kN}$

i.e. allowable load $= 8600\,\text{kN}$.

Adding $\frac{1}{3}(\gamma D A_b - W)$, the allowable load becomes $9200\,\text{kN}$.

According to the limit state method:

Characteristic undrained strength at base level, $c_{uk} = \dfrac{220}{1.50}\,\text{kN/m}^2$

Characteristic base resistance, $q_{bk} = 9c_{uk} = 9 \times \dfrac{220}{1.50} = 1320\,\text{kN/m}^2$

Characteristic shaft resistance, $q_{sk} = \dfrac{\beta\sigma'_0}{1.50} = \dfrac{86}{1.50} = 57\,\text{kN/m}^2$

Characteristic base and shaft resistances:

$$R_{bk} = \frac{\pi}{4} \times 3^2 \times 1320 = 9330\,\text{kN}$$

$$R_{sk} = \pi \times 1.05 \times 13.9 \times \frac{86}{1.50} = 2629\,\text{kN}$$

For a bored pile the partial factors are $\gamma_b = 1.60$ and $\gamma_s = 1.30$

Design bearing resistance, $R_{cd} = \dfrac{9330}{1.60} + \dfrac{2629}{1.30}$

$$= 5831 + 2022$$
$$= 7850\,\text{kN}$$

Adding $(\gamma D A_b - W)$ the design bearing resistance becomes $9650\,\text{kN}$.

## 8.12

(a)  $q_b = 9c_u = 9 \times 145 = 1305\,\text{kN/m}^2$

$q_s = \alpha\bar{c}_u = 0.40 \times 105 = 42\,\text{kN/m}^2$

For a single pile:

$Q_f = A_b q_b + A_s q_s$

$= \left(\dfrac{\pi}{4} \times 0.6^2 \times 1305\right) + (\pi \times 0.6 \times 15 \times 42)$

$= 369 + 1187 = 1556\,\text{kN}$

Assuming single pile failure and a group efficiency of 1, the ultimate load on the pile group is $(1556 \times 36) = 56\,016$ kN. The width of the group is 12.6 m, and hence the ultimate load, assuming block failure and taking the full undrained strength on the perimeter, is given by

$$(12.6^2 \times 1305) + (4 \times 12.6 \times 15 \times 105)$$
$$= 207\,180 + 79\,380 = 286\,560 \text{ kN}$$

(Even if the remoulded strength were used, there would be no likelihood of block failure.) Thus the load factor is $(56\,016/21\,000) = 2.7$.

(b) Design load, $F_{cd} = 15 + (6 \times 1.30) = 22.8$ MN

$$q_{bk} = 9c_{uk} = 9 \times \frac{220}{1.50} = 1320 \text{ kN/m}^2$$

$$q_{sk} = \alpha c_{uk} = 0.40 \times \frac{105}{1.50} = 28 \text{ kN/m}^2$$

$$R_{bk} = \frac{\pi}{4} \times 0.60^2 \times 1320 = 373 \text{ kN}$$

$$R_{sk} = \pi \times 0.60 \times 15 \times 28 = 791 \text{ kN}$$

$$R_{cd} = \frac{373}{1.60} + \frac{791}{1.30} = 233 + 608 = 841 \text{ kN}$$

Design bearing resistance of pile group, $\Sigma R_{cd} = 841 \times 36 \times 1.0 = 30\,276$ kN $= 30.27$ MN

$\Sigma R_{cd} > F_{cd}$ therefore the bearing resistance limit state is satisfied.

(c) Settlement is estimated using the equivalent raft concept. The equivalent raft is located 10 m ($\frac{2}{3} \times 15$ m) below the top of the piles and is 17.6 m wide (see Figure Q8.12). Assume that the load on the equivalent raft is spread at 2:1 to the underlying clay. Thus the pressure on the equivalent raft is

$$q = \frac{21\,000}{17.6^2} = 68 \text{ kN/m}^2$$

Immediate settlement:

$$\frac{H}{B} = \frac{15}{17.6} = 0.85$$

$$\frac{D}{B} = \frac{13}{17.6} = 0.74$$

$$\frac{L}{B} = 1$$

Hence from Figure 5.15:

$$\mu_0 = 0.78 \quad \text{and} \quad \mu_1 = 0.41$$

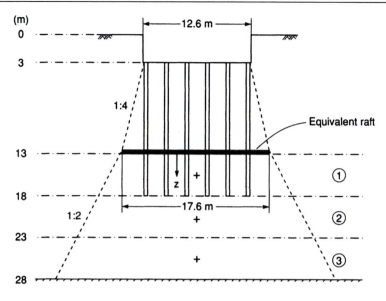

Figure Q8.12

Thus, using Equation 5.28:

$$s_i = 0.78 \times 0.41 \times 68 \times \frac{17.6}{65} = 6\,\text{mm}$$

Consolidation settlement:

| Layer | z (m) | Area (m²) | $\Delta\sigma$ (kN/m²) | $m_v\Delta\sigma H$ (mm) |
|---|---|---|---|---|
| 1 | 2.5 | 20.1² | 52.0 | 20.8 |
| 2 | 7.5 | 25.1² | 33.3 | 13.3 |
| 3 | 12.5 | 30.1² | 23.2 | 9.3 |
| | | | | $\overline{43.4\ (s_{od})}$ |

Equivalent diameter $= 19.86\,\text{m}$; thus $H/B = 15/19.86 = 0.76$. Now $A = 0.28$, hence from Figure 7.12, $\mu = 0.56$. Then, from Equation 7.10:

$$s_c = 0.56 \times 43.4 = 24\,\text{mm}$$

The total settlement is $(6 + 24) = 30\,\text{mm}$.

## 8.13

At base level, $N = 26$. Then using Equation 8.30:

$$q_b = 40N\frac{D_b}{B} = 40 \times 26 \times \frac{2}{0.25} = 8320\,\text{kN/m}^2$$

(Check: $<400N$, i.e. $400 \times 26 = 10\,400\,\text{kN/m}^2$)

Over the length embedded in sand:

$$\overline{N} = 21 \quad \left(\text{i.e.} \ \frac{18 + 24}{2}\right)$$

Using Equation 8.31:

$$q_s = 2\overline{N} = 2 \times 21 = 42 \, \text{kN/m}^2$$

For a single pile:

$$Q_f = A_b q_b + A_s q_s$$
$$= (0.25^2 \times 8320) + (4 \times 0.25 \times 2 \times 42)$$
$$= 520 + 84 = 604 \, \text{kN}$$

For the pile group, assuming a group efficiency of 1.2:

$$\sum Q_f = 1.2 \times 9 \times 604 = 6523 \, \text{kN}$$

Then the load factor is

$$F = \frac{6523}{2000 + 1000} = 2.1$$

(b) Design load, $F_{cd} = 2000 + (1000 \times 1.30) = 3300 \, \text{kN}$

Characteristic base resistance per unit area, $q_{bk} = \dfrac{8320}{1.50} = 5547 \, \text{kN/m}^2$

Characteristic shaft resistance per unit area, $q_{sk} = \dfrac{42}{1.50} = 28 \, \text{kN/m}^2$

Characteristic base and shaft resistances for a single pile:

$$R_{bk} = 0.25^2 \times 5547 = 347 \, \text{kN}$$
$$R_{sk} = 4 \times 0.25 \times 2 \times 28 = 56 \, \text{kN}$$

For a driven pile the partial factors are $\gamma_b = \gamma_s = 1.30$

Design bearing resistance, $R_{cd} = \dfrac{347}{1.30} + \dfrac{56}{1.30} = 310 \, \text{kN}$

For the pile group, $\sum R_{cd} = 1.2 \times 9 \times 310 = 3348 \, \text{kN}$

$\sum R_{cd} > F_{cd}$ (3348 > 3300), therefore the bearing resistance limit state is satisfied.

(c) Referring to Figure Q8.13, the equivalent raft is 2.42 m square.
For the serviceability limit state, the design load, $F_{cd} = 2000 + 1000 = 3000 \, \text{kN}$
The pressure on the equivalent raft, $q = 3000/2.42^2 = 512 \, \text{kN/m}^2$

From Figure 8.12, for $B = 2.42 \, \text{m}$, the value of $z_1$ is 1.8 m. Therefore $N$ values between depths of 1.33 and 3.13 m should be used. Thus

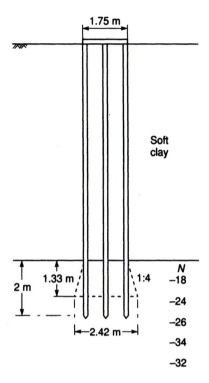

1.75 m

Soft
clay

2 m

1.33 m

1:4

2.42 m

N
−18

−24

−26

−34

−32

*Figure Q8.13*

$$N = \frac{24 + 26 + 34}{3} = 28$$

$$I_c = \frac{1.71}{28^{1.4}} = 0.016 \quad \text{(Equation 8.18)}$$

$$s = 512 \times 2.42^{0.7} \times 0.016 = 15\,\text{mm} \quad \text{(Equation 8.19(a))}$$

The settlement is less than 20 mm, therefore the serviceability limit state is satisfied.

**8.14**

Using Equation 8.41:

$$T_f = \pi D L \alpha c_u + \frac{\pi}{4}(D^2 - d^2)c_u N_c$$

$$= (\pi \times 0.2 \times 5 \times 0.6 \times 110) + \frac{\pi}{4}(0.2^2 - 0.1^2)110 \times 9$$

$$= 207 + 23 = 230\,\text{kN}$$

# Chapter 9

# Stability of slopes

## 9.1

Referring to Figure Q9.1:

$$W = 41.7 \times 19 = 792 \, \text{kN/m}$$

$$Q = 20 \times 2.8 = 56 \, \text{kN/m}$$

$$\text{Arc length, AB} = \frac{\pi}{180} \times 73 \times 9.0 = 11.5 \, \text{m}$$

$$\text{Arc length, BC} = \frac{\pi}{180} \times 28 \times 9.0 = 4.4 \, \text{m}$$

The factor of safety is given by

$$F = \frac{r\Sigma(c_u L_a)}{Wd_1 + Qd_2} = \frac{9.0[(30 \times 4.4) + (45 \times 11.5)]}{(792 \times 3.9) + (56 \times 7.4)} = 1.67$$

$$\text{Depth of tension crack, } z_0 = \frac{2c_u}{\gamma} = \frac{2 \times 20}{19} = 2.1 \, \text{m}$$

$$\text{Arc length, BD} = \frac{\pi}{180} \times 13\frac{1}{2} \times 9.0 = 2.1 \, \text{m}$$

$$F = \frac{9.0[(30 \times 2.1) + (45 \times 11.5)]}{(792 \times 3.9) + (56 \times 7.4)} = 1.49$$

The surcharge is a variable action, therefore a partial factor of 1.30 is applied. In the limit state method, the design values of undrained strength ($c_{ud}$) are (30/1.40) and (45/1.40) kN/m².

$$\text{Design resisting moment} = r\sum(c_{ud} L_a) = \frac{9.0}{1.4}[(30 \times 4.4) + (45 \times 11.5)] = 4175 \, \text{kN/m}$$

$$\text{Design disturbing moment} = Wd_1 + Qd_2 = (792 \times 3.9) + (56 \times 1.30 \times 7.4)$$

$$= 3628 \, \text{kN/m}$$

The design resisting moment is greater than the design disturbing moment, therefore overall stability is assured.

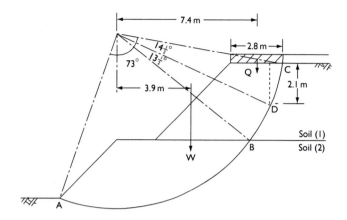

Figure Q9.1

## 9.2

$$\phi_u = 0$$

Depth factor, $D = \dfrac{11}{9} = 1.22$

Using Equation 9.2 with $F = 1.0$:

$$N_s = \frac{c_u}{F\gamma H} = \frac{30}{1.0 \times 19 \times 9} = 0.175$$

Hence from Figure 9.3:

$$\beta = 50°$$

For $F = 1.2$:

$$N_s = \frac{30}{1.2 \times 19 \times 9} = 0.146$$

$$\therefore \beta = 27°$$

## 9.3

Refer to Figure Q9.3:

| Slice No. | h cosα (m) | h sinα (m) | u/γw (m) | u (kN/m²) | l (m) | ul (kN/m) |
|-----------|-----------|-----------|----------|-----------|-------|-----------|
| 1 | 0.5 | – | 0.7 | 6.9 | 1.2 | 8 |
| 2 | 1.2 | –0.1 | 1.7 | 16.7 | 2.0 | 33 |
| 3 | 2.4 | – | 3.0 | 29.4 | 2.0 | 59 |
| 4 | 3.4 | 0.2 | 3.9 | 38.2 | 2.0 | 76 |
| 5 | 4.3 | 0.5 | 4.7 | 46.1 | 2.1 | 97 |
| 6 | 4.9 | 0.9 | 5.1 | 50.0 | 2.1 | 105 |
| 7 | 5.3 | 1.4 | 5.7 | 55.9 | 2.1 | 117 |
| 8 | 5.7 | 1.8 | 5.8 | 56.8 | 2.1 | 119 |

| Slice No. | $h \cos \alpha$ (m) | $h \sin \alpha$ (m) | $u/\gamma_w$ (m) | $u$ (kN/m²) | $l$ (m) | $ul$ (kN/m) |
|---|---|---|---|---|---|---|
| 9 | 5.9 | 2.4 | 5.9 | 57.8 | 2.2 | 127 |
| 10 | 5.9 | 2.9 | 6.0 | 58.8 | 2.2 | 129 |
| 11 | 5.6 | 3.3 | 5.7 | 55.9 | 2.3 | 129 |
| 12 | 5.2 | 3.5 | 5.2 | 51.0 | 2.4 | 122 |
| 13 | 4.6 | 3.7 | 4.5 | 44.1 | 2.5 | 110 |
| 14 | 3.4 | 3.2 | 3.4 | 33.3 | 2.7 | 90 |
| 15 | 1.6 | 1.9 | 1.8 | 17.6 | 2.9 | 51 |
|  | 59.9 | 25.6 |  |  | 32.8 | 1372 |

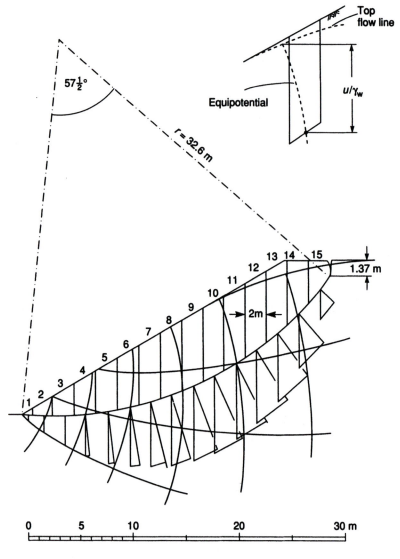

Figure Q9.3

$$\sum W \cos \alpha = \gamma b \sum h \cos \alpha = 21 \times 2 \times 59.9 = 2516 \, \text{kN/m}$$

$$\sum W \sin \alpha = \gamma b \sum h \sin \alpha = 21 \times 2 \times 25.6 = 1075 \, \text{kN/m}$$

$$\sum (W \cos \alpha - ul) = 2516 - 1372 = 1144 \, \text{kN/m}$$

$$\text{Arc length,} \ \ L_a = \frac{\pi}{180} \times 57\frac{1}{2} \times 32.6 = 32.7 \, \text{m}$$

The factor of safety is given by

$$F = \frac{c'L_a + \tan \phi' \Sigma(W \cos \alpha - ul)}{\Sigma W \sin \alpha}$$

$$= \frac{(8 \times 32.7) + (\tan 32° \times 1144)}{1075}$$

$$= 0.91$$

According to the limit state method:

$$\phi_d' = \tan^{-1}\left(\frac{\tan 32°}{1.25}\right) = 26.5°$$

$$c' = \frac{8}{1.60} = 5 \, \text{kN/m}^2$$

Design resisting moment $= (5 \times 32.7) + (\tan 26.5° \times 1144) = 734 \, \text{kN/m}$
Design disturbing moment $= 1075 \, \text{kN/m}$

The design resisting moment is less than the design disturbing moment, therefore a slip will occur.

**9.4**

$$F = \frac{1}{\Sigma W \sin \alpha} \sum \left[ \{c'b + (W - ub) \tan \phi'\} \frac{\sec \alpha}{1 + (\tan \alpha \tan \phi'/F)} \right]$$

$$c' = 8 \, \text{kN/m}^2$$

$$\phi' = 32°$$

$$c'b = 8 \times 2 = 16 \, \text{kN/m}$$

$$W = \gamma bh = 21 \times 2 \times h = 42h \, \text{kN/m}$$

Try $F = 1.00$

$$\frac{\tan \phi'}{F} = 0.625$$

Values of $u$ are as obtained in Figure Q9.3.

| Slice No. | $h$ (m) | $W = \gamma bh$ (kN/m) | $\alpha°$ | $W \sin \alpha$ (kN/m) | $ub$ (kN/m) | $c'b + (W - ub)$ $\times \tan \phi'$ (kN/m) | $\dfrac{\sec \alpha}{1 + (\tan \alpha \tan \phi')/F}$ | Product (kN/m) |
|---|---|---|---|---|---|---|---|---|
| 1 | 0.5 | 21 | $-6$ | $-2$ | 8 | 24 | 1.078 | 26 |
| 2 | 1.3 | 55 | $-3\frac{1}{2}$ | $-3$ | 33 | 30 | 1.042 | 31 |
| 3 | 2.4 | 101 | 0 | 0 | 59 | 42 | 1.000 | 42 |
| 4 | 3.4 | 143 | 4 | 10 | 76 | 58 | 0.960 | 56 |
| 5 | 4.3 | 181 | $7\frac{1}{2}$ | 24 | 92 | 72 | 0.931 | 67 |
| 6 | 5.0 | 210 | 11 | 40 | 100 | 85 | 0.907 | 77 |
| 7 | 5.5 | 231 | $14\frac{1}{2}$ | 58 | 112 | 90 | 0.889 | 80 |
| 8 | 6.0 | 252 | $18\frac{1}{2}$ | 80 | 114 | 102 | 0.874 | 89 |
| 9 | 6.3 | 265 | 22 | 99 | 116 | 109 | 0.861 | 94 |
| 10 | 6.5 | 273 | 26 | 120 | 118 | 113 | 0.854 | 97 |
| 11 | 6.5 | 273 | 30 | 136 | 112 | 117 | 0.850 | 99 |
| 12 | 6.3 | 265 | 34 | 148 | 102 | 118 | 0.847 | 100 |
| 13 | 5.9 | 248 | $38\frac{1}{2}$ | 154 | 88 | 116 | 0.853 | 99 |
| 14 | 4.6 | 193 | 43 | 132 | 67 | 95 | 0.862 | 82 |
| 15 | 2.5 | 105 | 48 | 78 | 35 | 59 | 0.882 | 52 |
| | | | | 1074 | | | | 1091 |

$$F = \frac{1091}{1074} = 1.02 \text{ (assumed value 1.00)}$$

Thus

$$F = 1.01$$

**9.5**

$$F = \frac{1}{\Sigma W \sin \alpha} \sum \left[ \{W(1 - r_u) \tan \phi'\} \frac{\sec \alpha}{1 + (\tan \alpha \tan \phi')/F} \right]$$

$$\phi' = 33°$$

$$r_u = 0.20$$

$$W = \gamma bh = 20 \times 5 \times h = 100h \, \text{kN/m}$$

$$(1 - r_u) \tan \phi' = 0.80 \, \tan 33° = 0.520$$

Try $F = 1.10$

$$\frac{\tan \phi'}{F} = \frac{\tan 33°}{1.10} = 0.590$$

Referring to Figure Q9.5:

| Slice No. | h (m) | W = γbh (kN/m) | α° | W sin α (kN/m) | W(1 − r_u) × tan φ' (kN/m) | sec α / 1 + (tan α tan φ')/F | Product (kN/m) |
|---|---|---|---|---|---|---|---|
| 1 | 1.5 | 75 | 4 | 5 | 20 | 0.963 | 19 |
| 2 | 3.1 | 310 | 9 | 48 | 161 | 0.926 | 149 |
| 3 | 4.5 | 450 | $15\frac{1}{2}$ | 120 | 234 | 0.892 | 209 |
| 4 | 5.3 | 530 | 21 | 190 | 276 | 0.873 | 241 |
| 5 | 6.0 | 600 | 28 | 282 | 312 | 0.862 | 269 |
| 6 | 5.0 | 500 | 35 | 287 | 260 | 0.864 | 225 |
| 7 | 3.4 | 340 | 43 | 232 | 177 | 0.882 | 156 |
| 8 | 1.4 | 28 | 49 | 21 | 3 | 0.908 | 3 |
| | | | | 1185 | | | 1271 |

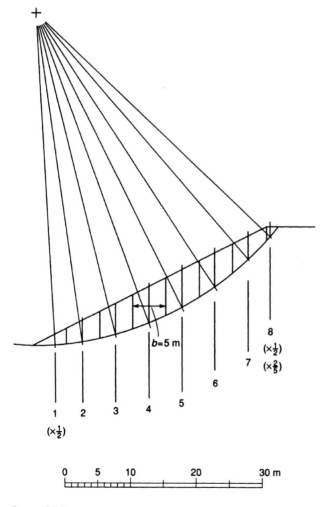

Figure Q9.5

$$F = \frac{1271}{1185} = 1.07$$

The trial value was 1.10, therefore take $F$ to be 1.08.

## 9.6

(a) Water table at surface; the factor of safety is given by Equation 9.12

$$F = \frac{\gamma' \tan \phi'}{\gamma_{sat} \tan \beta}$$

$$pti.e. \ 1.5 = \frac{9.2 \tan 36°}{19 \ \tan \beta}$$

$$\therefore \ \tan \beta = 0.234$$
$$\beta = 13°$$

Water table well below surface; the factor of safety is given by Equation 9.11

$$F = \frac{\tan \phi'}{\tan \beta}$$
$$= \frac{\tan 36°}{\tan 13°}$$
$$= 3.1$$

(b) $\phi'_d = \tan^{-1}\left(\dfrac{\tan 36°}{1.25}\right) = 30°$

Depth of potential failure surface $= z$

Design resisting moment per unit area, $R_d = (\sigma - u) \tan \phi'$
$$= \gamma' z \cos^2 \beta \tan \phi'_d$$
$$= 9.2 \times z \times \cos^2 13° \tan 30°$$
$$= 5.04 z \ kN$$

Design disturbing moment per unit area, $S_d = \gamma_{sat} \sin \beta \cos \beta$
$$= 19 \times z \times \sin 13° \cos 13°$$
$$= 4.16 z \ kN$$

$R_d > S_d$, therefore the limit state for overall stability is satisfied.

LaVergne, TN USA
03 August 2010
191780LV00003B/1/A